Applied Mineralogy
Technische Mineralogie

Edited by
Herausgegeben von

V. D. Fréchette, Alfred, N.Y.
H. Kirsch, Essen
L. B. Sand, Worcester, Mass.
F. Trojer, Leoben

4

Springer-Verlag
New York Wien 1973

E. M. Winkler

Stone: Properties, Durability in Man's Environment

Springer-Verlag

New York Wien 1973

Professor ERHARD M. WINKLER, University of Notre Dame, College of Science,
Department of Geology, Notre Dame, Indiana, U.S.A.

With 150 partly colored figures

ISBN 0-387-81071-4 Springer-Verlag New York-Wien
ISBN 3-211-81071-4 Springer-Verlag Wien-New York

To my wife Isolde

Foreword

As one of the most widely accessible building materials available to man, natural stone has been in extensive use for many centuries. It is a significant component, and in places the only one, of man-made structures the world over, and its properties, applications, and behavior over long periods of time constitute a story that is almost unbelievably complex. Important elements of the story are described and interrelated in this volume.

That the exposed parts of the earth's crust provide a considerable variety of rock types is evident to any thoughtful observer. To the geologist falls the task of characterizing and explaining this variety, but many other kinds of specialists who are involved in the commercial use of stone also have an essential stake in the matter. From quarryman to mason, from architect to structural engineer, and certainly from purchaser to future observer, there is compelling interest in the nature, appearance, and durability of one stone as compared with another, or of stone as compared with some other material. Small wonder, then, that much has been written on the subject, and that numerous aspects of commercial stone and its properties have appealed to a host of investigators. Research in this area also has been an official concern of many organizations, which in the United States include the American Society for Testing and Materials, the National Bureau of Standards, the U.S. Bureau of Mines, the U.S. Geological Survey, and several state agencies.

Despite an impressive and valuable body of accumulated data on occurrence, composition, physical characteristics, quarrying, preparation, installation, and maintenance of commercial stone, surprisingly little is yet known about the detailed responses of such natural materials to the various environments in which they have been used. To be sure, the geologist has long recognized that relatively little crustal material is in equilibrium with the physical and chemical conditions prevailing at the earth's surface, and he has learned to understand many of the changes involved in the weathering process. But weathering of undisturbed rocks at the outcrop often is little more than a crude indication of what can occur in stone that has been detached, perhaps dimensioned and finished, and moved to a new site. The nature and rate of attacking processes can change drastically, especially if the new site is in a relatively corrosive urban or industrial environment.

The observable results of exposure in use, though salutary for a very few kinds of rocks, are ordinarily less than desirable. They range from roughening of polished surfaces to large-scale cracking and spalling, and from solution pitting to staining and other kinds of disfigurement. Carefully selected and prepared stone can provide important advantages over most other kinds of building and monumental materials, notably in the areas of appearance and durability, but it nonetheless may deteriorate appreciably in long-term use. He who doubts this need only examine some of the major structures in any large city, the markers in an old cemetery, or the Washington Monument!

In this book, Professor WINKLER has addressed himself to the difficult task of appraising the stability and durability of stone as it has been used in various ways by man, and of relating its behavior in a fundamental manner to the various attacking materials and processes found in contrasting environments. He has carefully examined and analyzed a published record that is extensive but highly dispersed, and he has found that in general it reflects two principal kinds of investigations — those with broad scope and an empirical approach, and those aimed at more fundamental solutions of specialized problems. He also has drawn upon his own widespread observations and researches in an effort to provide insights into pertinent variables and to develop an improved understanding of the entire problem. His presentation brings us up to date in an important area not heretofore so considered in one volume, and he offers a series of suggestions for improved conservation of stone. He also leaves us with some notion of how much remains to be learned!

It seems appropriate to confirm here what is implicit in the pages of this book. Professor WINKLER is not merely a well-informed scientist; he also has lithic interests that border on genuine personal affection for commercial stone. He deeply enjoys the look and the feel of a polished granite panel, just as he suffers a bit from the sight of a rusty, corroded cornice. To him the facing of a structure can be the earth's crust in microcosm, and when he visits a cemetery, notebook in hand, he reads both the inscriptions and the rocks. For all this I offer him a sentimental salute, along with a more pragmatic one for his courage in preparing a book on so complex a subject.

Stanford University,
Stanford, California, October, 1972   RICHARD H. JAHNS

Preface

Stone is the primary building material of the earth's crust; in its basic function it has appealed to man's most primitive needs and has stimulated his artistic sense since the dawn of civilization. Obelisks, pyramids and colossal stone sculptures of early cultures reached astonishing perfection both in excellence of workmanship and in technique of stone transport.

Despite its antiquity, stone is regaining great popularity as a building material through a revolution in the art of quarrying and finishing. Unlimited combinations of stone textures and colors offer the architect a wide range of applications. Stone: Properties and Durability in Man's Environment, surveys the scientific principles regarding the important stone properties pertinent to the architect's, the engineer's and the stone producer's needs.

Our attention to lasting stone beauty and proper maintenance has increased in recent years in proportion to the acceleration in visible damage to stone exposed to polluted urban air and waters. The time-lapse pictures on page 87 underscore this theme: they show, since the beginning of industrialization, near-exponential acceleration of stone decay on a statue in the heavily industrialized Rhein-Ruhr area of northwestern Germany. The study of stone in its manifold interactions with a complex and hostile environment can only be understood if all components are explored, including the nature of weathering agents themselves. The information is as applicable to concrete aggregate stone as to building stones and ornamental stones.

Stone weathering has long fascinated this author, who has made many observations on stone decay, especially in the immediate vicinity of the University of Notre Dame campus. Field trips with students have inspired the author to assemble and organize the available information on stone with respect to its properties, durability and use. Few readers need travel very far to see damage inflicted to stone monuments similar to many of the cases cited and illustrated in this book.

Graphs and diagrams, the language of precise science, are used to enable quantitative treatment of processes involved in stone decay. These graphs, mostly selected from the scientific literature, have been modified for greater clarity to the non-scientist. Brief summaries conclude each chapter for better understanding of the subject to persons lacking a scientific background.

Hope is expressed that this book will be valuable to all those persons who deal with natural stone. Costly mistakes in the selection of material should be avoidable in the future through a better understanding of stone as a very heterogeneous substance whose interaction with the complex industrial and urban environment must be painstakingly evaluated. The reader will note that many important questions still await answers.

This book should appeal to readers with a variety of interests and backgrounds, such as:

the stone producers, for information on properties and durability of his product;

the architect, planner and engineer, for color, texture and durability;

the student of geology, for physical stone properties and stone weathering in both rural and urban environments;

the engineer, for weathering, strength of aggregate stone and the environment of facing stone, rip-rap, concrete;

the ecologist, for general information on our urban environment, and

the conservator, for modern practices in the preservation of stone buildings and monuments.

The preparation of this text has been greatly assisted by Professor V. D. FRÉCHETTE, the Springer Applied Mineralogy Series Editor, with many valuable suggestions and corrections, by Professor RICHARD JAHNS, Dean of Earth Science at Stanford University, for his apt and valuable foreword, and by Dr. SCHMIDT-THOMSEN, Landesdenkmalamt Westfalen-Lippe, Germany, who supplied important documents such as time-lapse photos of decaying monuments; to them I wish to express my sincere gratitude. Above all I am grateful to my wife ISOLDE for continuous assistance, encouragement, suggestions and great patience.

Notre Dame, Indiana, in Fall 1972   ERHARD M. WINKLER

Table of Contents

1. Rock and Stone

Rock is the basic building material of the earth's crust, and the original building material used by man. All rocks, called stone if fabricated, are composed of one or several kinds of minerals — these help to determine the physical and chemical properties of rocks. Detailed descriptive information on mineral properties is omitted here; it may be readily obtained from any text book on physical geology. The information, however, is summarized on a chart in Appendix A, with all mineral properties pertinent to the stone industry. Rocks are classified into three major groups based on their origin, igneous, sedimentary and metamorphic rocks:

1. Igneous or magmatic rocks are primarily crystallized from a fiery fluid silicate melt, either deep below the earth's surface or at the surface. Texture and fabric of these rocks depend on their environment during crystallization. Granite, gabbro, basalt, porphyry, and others belong to this group.

2. Sedimentary rocks, or layered rocks, are formed by the concentration of debris of variable size and shape deposited by mechanical means or by precipitation, or by accumulation of organic skeletons and shells. Conglomerate, sandstone, shale, limestone marble, dolomite, travertine, and onyx marble are common sedimentary rocks.

3. Metamorphic rocks are derived either from igneous or sedimentary rocks recrystallized by the effect of pressure and temperature. Important rocks of this group are gneiss, slate, marble and crystalline quartzite. Figure 73 of the chapter Decay of Stone sketches the environments of the three main rock groups.

Mineral composition, fabric, texture, structure and color, are important characteristics of rocks.

Mineral composition: Rock properties depend on the physical and chemical characteristics of the minerals to a large extent; the coarser the mineral constituents, the better the individual components will stand out and show their differential behaviour.

Fabric: The rock fabric is the spatial orientation of mineral, or gross rock components, whether crystals or fragments.

Texture: Rock texture is the geometrical aspect of the component particles, including size, shape and arrangement. Texture overlaps with fabric; for the sake of clarity most geology texts combine texture and fabric to "texture". Texture characterizes the grain sizes, shapes and grain contact. Textures differ greatly among the three major rock groups.

The structure of a rock mass consists of the larger features such as bedding, flow banding, cross bedding and others. In a wider sense structure includes fractures, cleavage, brecciation, folding, feather jointing, etc. Major structural features are treated in a separate chapter.

1.1. Igneous Rocks

Igneous rocks are also called "primary rocks" because they crystallize from a hot original silicate melt, the magma. If cooling progresses very slowly beneath the crust, crystallization is slow and the resulting crystals are coarse grained. Rocks resulting from this environment of formation are granite, syenite, diorite

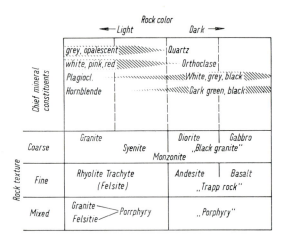

Fig. 1. Classification of igneous rocks, simplified

and gabbro. If cooling takes place rapidly at or near the earth's surface, very fine grained felsite, basalt or porphyry will result. The classification of the igneous rocks is mainly based upon the mineral content: coarse-grained (phaneritic) intrusive rocks, and fine-grained (aphanitic) extrusive rocks with a dense appearance (Fig. 1).

Igneous Textures

refer to the size and shape of the individual mineral grains.

a) *Phaneritic:* almost all minerals crystallized slowly under nearly the same conditions with an appearance of almost equal grain sizes. Quartz grains and mica flakes are, as a rule, much smaller than the prismatic feldspar and hornblende.

b) *Aphanitic or non-visible textures:* rapid cooling at or very near the earth's surface crystallized the entire rock mass homogeneously. The minerals and their textural arrangement can be revealed only in a rock thin section. Felsite, basalt, and some porphyries are the most important representatives. As a rule, finer grained rocks are harder than their coarser grained equivalents.

c) *Porphyritic texture:* a mixture of coarse and fine grained minerals. Coarse constituents with well developed crystal outlines float in a distinctly finer groundmass or matrix. The contrast of the different crystal sizes reflects different environments of crystallization, i. e., the larger crystals probably crystallized first deep below the crust, the finer grained matrix near or at the surface later on. Porphyritic granites are generally attractive as architectural and monumental stone (Fig. 2).

Fig. 2. Porphyritic granite from Marble Falls, Texas, with large, well-defined phenocrysts of light feldspar. Walk is of grey sandstone; flaking of thin layers is evident. YMCA Building, Milwaukee, Wisconsin

Fig. 3. Delano granite (Minnesota) with pegmatite dike, showing gradational transition at the contact. Flame-surface finish of granite brings out natural appearance, permits full reflection from cleavage planes. Time-Life Building, Chicago

d) *Pegmatitic texture:* Crystals may be exceptionally large along veins and dikes. Although very ornamental, such dikes create a twofold problem: cleavage of the large feldspar and mica crystals, and easy separation of the dike rock from the wall rock if the contact is sharp. A gradation of the large sized dike to the finer grained rock generally eliminates the possibility of separation. Pegmatite veins are therefore considered as flaws by the stone producer for certain applications, but they are very attractive to the architect (Fig. 3).

Fig. 4. Minnesota granite (Cold Spring) with pegmatitic vein (marked with arrows). Sharp contact provides waterway for moisture. Slab for steps is about $1\frac{1}{4}''$ thick. Notre Dame, Indiana

e) *Aplitic texture:* Light-colored, sugar-grained rock in veins is often found as narrow dikes in generally coarser grained wallrock. Aplitic rock is often confused with white marble as aplites lack dark minerals. Both pegmatites and aplites are the result of a late magmatic phase. Sharp contacts to pegmatite and aplite dikes can be a source of water seepage (Fig. 4).

Igneous Structures

Intrusive bodies are complex structures formed deep below the earth's surface where moderate pressure is present and loss of temperature is very slow. Convectional flow of the semi-liquid magma and interaction with the wallrock may also be involved. Marginal zones of large intrusive bodies (batholiths) usually show such features if exposed. An idealized intrusive body is pictured in Fig. 5 as a block diagram in which a thick roof zone was removed by erosion. The diagram distinguishes between the massive core in the center with random orientation of the minerals, and flowfolding along the edge of the core resembling metamorphic rock with frequent curling and twisting of the magma by movement during the late intrusive phase, exemplified by the "Rainbow Granite" of Minnesota (Fig. 6). True metamorphic rocks occur towards the outer margin of an intrusive mass. Such rock should be carefully tested for strength along the flow structure, or

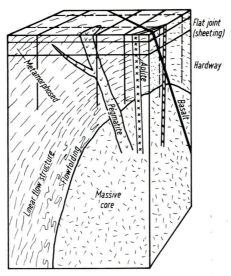

Fig. 5. Structure of igneous rocks

Fig. 6. Rainbow-Granite from Cold Spring, Minnesota, used as exterior store-front veneer. Flowbanding around coarse-grained orthoclase crystals in the center with light and dark bands alternating

foliation. A mere blow with the hammer against the block in question determines if sufficient strength will keep the rock together and prevent damage by salt and freezing. Loose mineral bonding or the presence of minerals with perfect cleavage or parallel mica flakes may render the stone worthless for dimension stone, but may produce good flagstone for flooring.

Classification of Igneous Rocks

Igneous rocks range from almost white granites to black gabbros and basalts. Acid magma, high in silica, crystallizes to rocks which are generally high in orthoclase, quartz, but low in dark minerals like hornblende and black mica. In contrast, relatively basic magmas, low in silica, crystallizes to dark colored rocks which are high in grey-to-black plagioclase and hornblende. The color of the igneous rocks depends mainly on the color of the prevailing feldspars which generally make up between 50 and 75% of the rock substance. Feldspar colors of granites and syenite may be white, pink, flesh or deep red, whereas diorites and gabbros

Fig. 7. Reddish-brown scoria basalt, shaped into rectangles of precise dimensions and set without cement, a modern version of ancient Aztec, Maya, and Inca technique. Mexico City, Mexico

are medium grey to black. The quartz in granite may amount to less than 10%, to a maximum of about 35%; its percentage depends on the abundance of silica in the magma during crystallization. The danger of silicosis and a decrease of fire resistance should be expected in granitic rocks with increasing quartz content. Syenites are quartz-free granites with a higher per cent of dark constituents. The rock classification chart of the igneous rocks in Figure 1 is based upon both the mineral composition and the texture. The commercial stone names are placed in quotation marks where they are different from the scientific terms. Granite is the most frequently used igneous rock because of its abundance and great variety of color and textures. Diorites and gabbros, often advertised as "black granites" are less frequently seen. Slag-like basaltic scoria are marketed as slag stones for exterior decoration. Some slag basalts are filled with mineral matter of different color, mostly white or light green. The basalt stones of Mexico City should be mentioned here as an impressive example of rusty-red lava, often with deep green or white cavity fillings and often elongated (Fig. 7). Pale red or olive green porphyries with large, well-defined crystals attracted the ancient Egyptian and Greek sculptors. The same stone is marketed today under a variety of names, e.g., Porfido Rosso, Porfido Verde Antico, etc. Volcanic fragmentary rocks, such as ashes and cinders, are not useful.

1.2. Sedimentary Rocks

Sedimentary rocks, or layered rocks, are formed either by the accumulation of rock and soil material by streams, waves or wind, or as organic accumulations and chemical precipitates. A flat, layered structure is an important characteristic of sedimentary deposits, known as bedding or stratification. The physical properties of sedimentary rocks depend primarily on the mineral composition, the texture, fabric, structure, and, in the case of clastic sediments, on the cement between the fragments. Sedimentary rocks are generally classified as

a) clastic sediments, coarse to fine grained: residual rock fragments produced through rock disintegration, result in various sizes and shapes;

b) clastic sediments, fine to submicroscopic: residual silt and clay-sized particles form the weathering end product of mostly feldspars and some quartz;

c) organic sediments: the remains or products of animals and plants: many limestones, limestone-marbles and dolomites belong to this group of rocks;

d) chemical precipitates: precipitation from ocean waters and brines lead to some limestones, dolostones (dolomites), gypsumrock, saltrock.

A mixture of these sedimentary materials is common in nature as many natural environments of deposition overlap. A simplified genetic relationship of sedimentary rocks is presented in Table 1.

Table 1. *Classification of Sedimentary Rocks*

A. CLASTIC	B. CHEMICAL BIOCHEMICAL	C. ORGANIC BIOCHEMICAL
Residues soils	Transitional calc. shale	Shell limestone coral limestone, coquina	Org. Residue coal, oil
Washed Residue conglomerate, sandstone, siltstone, shale	PRECIPITATED SEDIMENTS limestone, dolostone, chert, gypsum rock, saltrock	

Modified from: PETTIJOHN (1957).

1.2.1. Clastic Sediments

Mineral Composition

Most minerals in clastic sediments were inherited from primary igneous rocks or from previously existing sedimentary or metamorphic rocks. Quartz survives mechanical abrasion in grinding processes in stream beds or on beaches, as the hardest common mineral which is also resistant to chemical weathering. Carbonate fragments are attacked by transporting waters, which waters are generally lime-hungry. Feldspars and ferro-magnesium silicates slowly weather to clay. Quartz becomes then singled out as the sole survivor of the process of mass transportation and weathering. Soft, elastic mica flakes may survive chemical attack and some abrasion during transportation to join quartz and clay minerals as a residue.

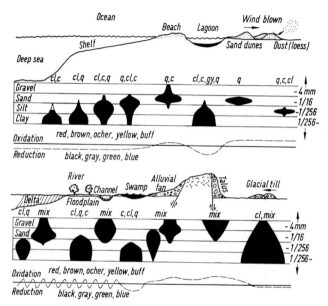

Fig. 8. Sedimentary environments: illustration of grain size distribution, approximate mineral composition, and degree of oxidizing or reducing environment. Adapted from FOLK (1957) Legend for abbreviations: c=calcite, cl = clay minerals, q=quartz, gy=gypsum, mix=mixture of minerals

Clastic sedimentary textures consist of the following parameters:

a) *Sorting* indicates the degree of similarity of grain sizes which reflect the transporting agent. Fig. 8 gives a graphic presentation of the genetic environment of clastic sediments. The graph may serve as a crude model for grain size distribution, mineral composition and expected colors of sediments. The unsorted glacial till compiled by snowplow action of an advancing ice sheet contrasts with a very well sorted dune sand compiled by very selective wind action. Sand may appear as part of the rock matrix between larger fragments of breccias or conglomerates. The matrix should not be confused with the rock cement. The same kind of deposit may expose quite different grain sizes within a very narrow space;

0.9
0.7
0.5
0.3
Decreasing Sphericity
0.1
0.3
0.5
0.7
0.9
Decreasing Roundness

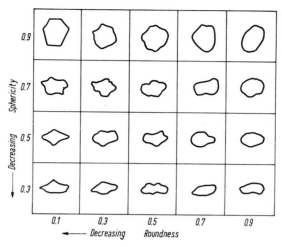

Fig. 9. Relationship of sphericity and roundness. After BAYLY (1968). Sphericity is related to the various diameters of the particles while roundness is related to the sharpness of the corners

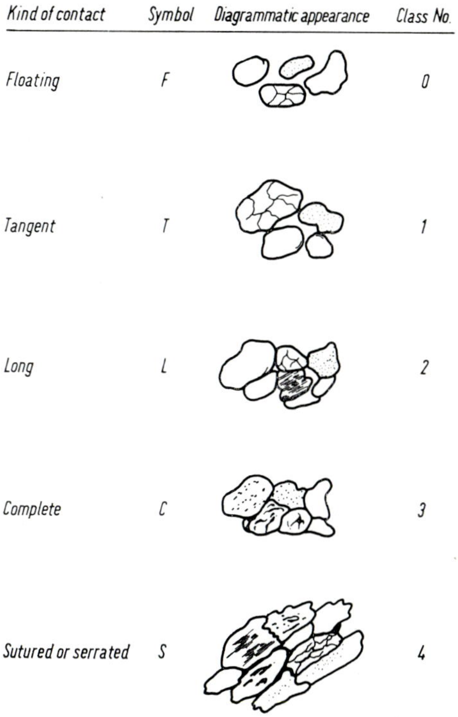

Kind of contact
Symbol
Diagrammatic appearance
Class No.
Floating
F
0
Tangent
T
1
Long
L
2
Complete
C
3
Sutured or serrated
S
4

Fig. 10. Grain packing. After GRIFFITHS (1967)

alluvial fans often alternate coarse flood material with fine sand and silt deposited during minimum water flow; the same should be expected with sediments of deltas. Oxidizing or reducing conditions during the time of sedimentation determine the original color, i.e., red, brown, yellow and ochre under oxidizing conditions, and black, grey, green and bluish-green under teducing conditions. The original color of the sediment may change, however, during later exposure dowing to a change in conditions. Some red limestone marbles were exposed to excessive oxidation during tectonic crushing in the geologic past.

b) *Rounding* expresses the ratio of the average radius of curvature of several edges of a solid to the radius of curvature of the maximum inscribed sphere. KRUMBEIN's visual roundness scale is used where rapid identification is desired (Fig. 9).

c) *Sphericity* is the degree of rounding in which the shape of a fragment approaches the form of a sphere. By means of visual comparison two-dimensional sphericity may range from angular to perfectly round.

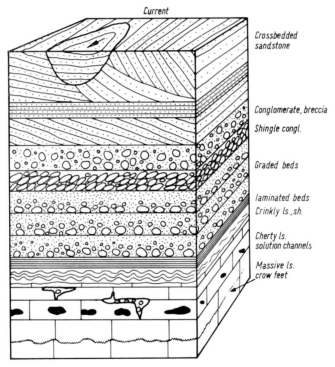

Fig. 11. Structures of sedimentary rocks
Legend for abbreviations: *ls*=limestone, *sh*=shale

d) *Packing* of the sedimentary grains is the nearness of the grains to one another, presented by GRIFFITHS (1967) as the configuration of the grain contacts. Fig. 10 distinguishes between floating, tangent, long, complete and sutured or serrated contacts; the terms refer to the relationship of the grains to the intergranular spaces and spacings. "Serrated" mineral contacts are generally only found in metamorphic or near-metamorphic rocks.

e) *Fabric* of sediments is the orientation — or the lack of orientation — of the rock fragments. The orientation may be random (isotropic), or oriented to a quasi-shingled arrangement by swiftly running water or strong wind (Fig. 11).

Cement of Clastic Sedimentary Rocks

Dissolved salts travel into pores of clastic sediments by groundwater or capillary action where they precipitate between the fragments as natural cement. Three different degrees of cementation are distinguished:

a) *Contact cement:* A thin film of precipitated mineral matter coats the individual grains and cements the grains against each other at the points of contact. The porosity of a contact-cemented clastic rock is quite high unless the grain packing is serrated. The rock strength may be satisfactory with contact cement provided only that the cementation is not spotty and that the cement is calcitic or quartzose.

b) *Pore cement* fills the interstices between the grains whether or not contact cement had previously coated the fragments.

c) *Basal cement:* Pore cement may be called basal cement if the cement occupies a large volume of the rock in floating packing of the grains. The rock strength depends on the kind of cement and the degree of cementation.

Mineral Composition of Clastic Rock Cement

Strength and durability of the rock depend on both the composition of the cement and on the degree of cementation. The following important mineral cements are recognized in commercial stone:

a) *Siliceous cement:* Silica may occur as finely crystalline quartz, as microcrystalline or cryptocrystalline chert or chalcedony, or as amorphous opal. Rocks cemented with silica may reach very high strength provided that all pores are filled with cement and no clay film coats the fragments: The rock should break across the grains rather than around them if the cement has grown in crystallographic continuity with the fragments; the term orthoquartzite is used whereby the prefix "ortho" is often omitted. Sedimentary quartzites are readily distinguished from true metamorphic quartzites by their lack of uniform crystallinity.

b) *Carbonate cement:* Calcite cement is the dominant cement in sedimentary rocks. Thorough cementation with calcite or dolomite presents a rock of sufficient strength and durability for most technical applications.

c) *Ferric oxide and ferric hydroxide:* Iron oxides are occasionally introduced as both contact or pore cement or both. Cementation with hematite or limonite is usually incomplete, the rock strength is low and its resistance to weathering poor.

d) *Clayey cement:* Clay as contact cement around larger fragments is more common than generally believed; its presence, however, is difficult to detect with petrographic methods. Clay contact cement may be quite sound in dry climate; disintegration in humid climate is rapid both by swelling action of the clay minerals, and by frost action and salt burst. Clay cement can be introduced as mud which coated the gravel and sand fragments in the river bed or on beaches which dried and adhered during the process of burial.

Structure of Clastic Sediments

Sedimentary internal rock structures refer to the kind and thickness of bedding. Fig. 12 combines the scientific with a practical classification of thickness of bedding which was observed to depend much on both the clay and the silica content. Silica in sediments increases the thickness at constant lime content, whereas clay diminishes bedding from massive towards laminated.

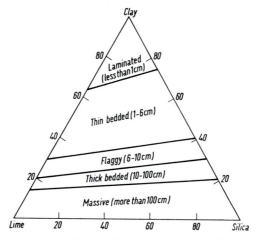

Fig. 12. Bedding classification as influenced by the proportions of lime to silica to clay content. After PETTIJOHN (1957), also modified from data of EICHER (1968)

Fig. 13. Crossbedded sandstone. Light streaks are green zones of ferrous iron formed by reduction of portions of the red matrix. Saw markings on stone surface are visible. Pencil rests in recessed building joint. Frankfurt, Germany

    a) *Massive banks* of sandstone and conglomerate exceed a thickness of 100 cm. (3 ft.). Sandstones in this category are often crossbedded. The cement for this thickness is siliceous or calcitic.

b) *Flags*, or flaggy beds are of medium thickness and break into "flags" of about 6 to 10 cm. thickness. Sandstone flags are common and useful for walks and patios if strong enough between the partings. The walking surface is usually sufficiently granular to be slip-proof when wet; wavy surfaces from fossil ripple marks and flow markings add character to the stone.

c) *Laminated beds*, are less than 1 cm. thick, too thin to serve as flags.

d) *Crossbedding* is composed of a sequence of horizontal strata alternating with uniformly inclined beds. The term crossbedding may mislead, as criss-cross beds are barely found in nature. The direction of dip of the inclined beds (foreset beds) is quite consistent with the direction of the former current as should be expected in sanddunes (Fig. 13) and small deltas. Channel fillings are common in old stream deposits where an interruption of the bedding continuity can develop as cut-and-fill structures. Partings along crossbeds occur as frequently as along horizontal beds; occasional crumbling should therefore be expected also from inclined crossbeds (Fig. 14).

Fig. 14. Crossbedded Indiana Limestone starts crumbling along crossbeds where inclined beds dip outward. Post Office Building, South Bend, Indiana

e) *Graded bedding* is a type of stratification in which each bed displays gradation of grain size from coarse at the bottom to fine at top. This sequence repeats itself in each bed.

f) *Shingle structure* is a distinct, cross-bed like arrangement of flat pebbles of various colors, like roof shingles. Though rare, shingle conglomerates give a distinct and directional pattern to a dimension stone.

Grain Size Classification

Clastic sediments are classified on the basis of their mean grain size as follows:

1. **Conglomerate** is a rock which is composed of well worn and well rounded pebbles of various sizes which may be cemented together to a more or less strong rock, often resembling concrete. The pebbles may also float in a matrix of much

finer granules and sand. Differences in stream flow result in such mixtures. Great distances of transport usually mix together fragments of different colors and composition, for which reason conglomerates are often used as decorative stone. The degree of resistance to weathering for the various fragments shows up after a few years of exposure to the atmosphere especially where carbonate rocks are mixed with silicate rocks: loss of polish and pockmarking of the stone surface are usually the result.

2. Breccia or Breche is a coarse — to medium—grained fragmentary rock with angular, unworn fragments which are generally of uniform composition and color. Angular breches in large varieties are often seen as decorative stone (Fig. 15).

Fig. 15. Pieta statue of fine-grained, massive Carrara marble. Stand of monument is veneer of marble breccia. Orientation of some fragments by post-breccia metamorphism. Fine-grained matrix between angular white marble fragments is also slightly metamorphosed. "Pieta" by Mestrovic, Notre Dame, Indiana

3. Sandstone is composed of sand grains, mostly quartz, some feldspar, but also calcite, with a variety of grain shapes and grain cements (Fig. 14). A few commercial sandstone varieties are discussed in the following:

a) *Quartzite* is a quartz sandstone or conglomerate that is predominantly composed of quartz; its cementation with silica is so complete that the rock usually breaks across the grains instead of around the grains. Recrystallized metamorphosed quartz sandstone may also be called a quartzite. Both quartzites are very dense and exceptionally hard.

b) *Bluestone* is a hard, indurated feldspathic grey sandstone which splits easily into thin slabs. The stone is usually dark grey, rarely bluish.

c) *Brownstone* is a sandstone of distinctly brown or reddish-brown color, which has become a very popular building stone in the Eastern USA despite its limited durability.

d) *Freestone* is a sandstone which splits with equal ease into any desired direction and dresses easily due to incomplete cementation of the sand grains.

1.2.2. Chemical Sediments and Evaporites

The classification chart of sediments in Table 1 does not draw sharp lines between chemical sediments, evaporites and biochemical sediments, as one type

Table 2. *Nomenclature of Limestone-Shale Mixtures*

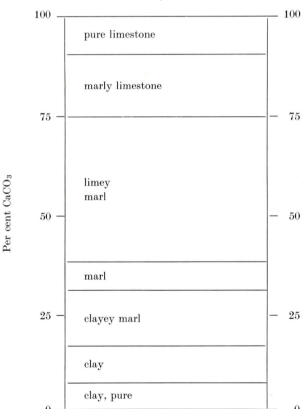

Modified from: PETTIJOHN (1957).

grades into another. Limestones and dolomites are often impure, with gradation to pure shale; a classification of common clay-lime mixtures was established by PETTIJOHN (1957) and is reproduced in Table 2.

Textures of Chemical Sediments

a) *Fine grained:* the grains are mostly invisible to the naked eye; the rock may be so fine grained that the fracture appears curved. Very fine-grained limestones were occasionally used for lithoprinting; the term lithographic limestone is then appropriate.

b) *Fragmentary:* some limestones may be composed of broken skeletal parts of animals and plants, but may also be brecciated by tectonic processes. Rounding of such fragments by intra-formational solution and subsequent recementation may cause a nodular texture. Complete oxidation during the process of fragmentation can change the original grey or buff colors from pastel buff to deep red. The Verona Red of Italy and probably some red Austrian limestone-marbles should be named here.

c) *Crystalline:* a coarsely crystalline limestone or gypsum rock recrystallized during diagenesis to larger grain size; this is a first step towards a true metamorphic crystalline marble. The surfaces of calcite cleavage planes are often plates

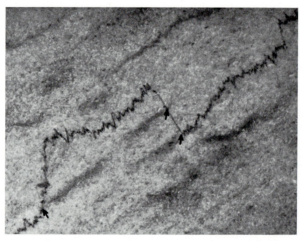

Fig. 16. Stylolites, often called "crow feet" in the stone industry, parallel to bedding plane of crystalline Holston marble, Tennessee. Note interruption of bedding plane by stylolite seam. Point of offset is marked with arrows

of sea-lily stems, starfish plates and other organic forms. A crystalline limestone resembles a metamorphic marble except for the presence of undistorted fossil shells, warm iron colors instead of cold whites or greys; the tan to pink Holston Marble of Tennessee is a popular stone of this category (Fig. 16).

d) *Oölitic,* or roe-shaped and roe-sized limestones often show a finely crystalline habit with ooids interspersed with concentric shells of calcite which were deposited in shallow, warm water. Oölitic textures may be sometimes discernible with the unaided eye, more often the microscope is required. Some varieties of the Indiana Limestone are oölitic.

Structures of Chemical Sediments

a) *Bedding* characterizes also chemical sediments. The lack of clay eliminates thin lamination. Thickness of bedding has been discussed in connection with clastic sediments.

b) *Crinkly beds* are generally limestones with occasional crenulation, found with some thin-bedded evaporite limestones.

c) *Ripple marks* and *mudcracks* may be observed on surfaces of fine-grained clastics and carbonates.

d) *Crow's feet (stylolites)* are irregular wavy seams which resemble graphs of daily temperatures. Stylolites are probably the result of interstratal pressure-solution in carbonate rocks during which process solution and crushing of grains cause the upper strata to collapse into the lower undissolved bed, filling the gap

Fig. 17. Exterior wall of dolomite rubble. Flat pieces with smooth surface are broken across whereas larger, rough-surfaced blocks split along stylolite planes. The key bunch shows the scale

with the insoluble, residual clay (WEYL, 1959). The most common stylolites run approximately parallel to the bedding planes of limestones as preferred diffusion and ion removal occurs along clay partings; occasionally stylolites may cross-cut bedding planes, even run vertical, or may form an interconnected network. PARK and SCHOT (1968) discuss the different types and ideas on stylolite formation. Various stages of formation are apparently involved. Residual concentration of organic matter and clay often thicken near the apex of the zig-zag lines (see Figs. 16, 17 also 22). The clay seams are lines of weakness in the stone although quite ornamental in appearance. Thick clay fills do not polish, and weather out first.

Types of Chemical Sediments and Evaporites

a) *Limestone* is principally composed of calcium carbonate. Clay impurities should be less than 5% (see Table 2). Limestones are subject to large variations as to lithification and composition. Limestone varieties low in clay take a good polish and are durable; they qualify for commercial limestone-marble.

b) *Magnesian (dolomitic) limestone:* A limestone which contains not less than 5% but more than 40% of magnesium carbonate. Many commercial limestones and limestone-marbles fall into this category.

c) *Dolomite* is a carbonate rock which contains calcium carbonate but more than 40% magnesium carbonate. High-magnesian dolomites are often finely crystalline and porous due to extensive recrystallization of the original low-magnesian carbonate rock. Dolomites are generally harder, denser and more brittle than limestones. Their use is the same as that for limestones.

Fig. 18. Travertine from Utah, coarse-grained and very porous, with deep-brown bands alternating with light-tan bands. Interior hallway wall of office building in Washington, D.C.

d) *Travertine* is a variety of banded limestone regarded as a product of chemical precipitation from cold lakes or streams, also hot springs. The loose, bedded and often porous structure creates marked heterogeneities. Colors of commercial travertines range from light cream and brown to maroon. Dense varieties with contrasting beds (bands) are favorites as exterior decorative panels. Fine-textured Italian travertine, the stone of ancient Rome, is seen most; colored varieties with coarser bedding and stronger contrast in color and texture are found in Colorado, Utah and Idaho travertines (Fig. 18).

e) *Onyx-marble* is a spring deposit of cold or hot waters which has recrystallized to a translucent rock of very fine grain size. Pastel colors of off-white, ivory, light green, golden brown and others make excellent translucent walls and window panels spreading warm light inside. Well-developed natural fractures in the stone, associated with the former penetration of ferric oxides of various concentrations and color shades, require much caution in milling and handling. The refraction and reflection of light in onyx panels for bathroom walls and floors gives almost a glass-opal-like effect. Onyx marbles from various locations in Mexico, Italy and Iran are most frequently used, recently for translucent panels as a substitute for windows or entire walls where very beautiful light effects are provided inside buildings during the day and, by inside lighting, outside at night.

f) *Shell limestones* contain numerous fossil sea shells of various size and shapes. Well-cemented and properly oriented, the shells show up as cross sections after fabrication of the rock into slabs. Stong contrasts are brought out between white shell cross sections and deep black limestone, for instance in the Blue Belgian Marble. The shell imprints of the buff Austin Shell Stone of Texas is another type of shell stone with many large vugs where the original shell substance was removed by selective solution. Many examples of commercial shell limestone and limestone-marble can be included here.

g) *Coral limestones* often comprise entire coral colonies and reefs on a single stone slab with often white coral branches against a grey or dark matrix. Intersections of such colonies with stone surfaces present a vivaceous pattern with rapid changes of the fabric. Cross sections through globular colonies are especially striking. The Key Largo Limestone of Florida and some older Alpine commercial limestone-marbles are examples (Fig. 19).

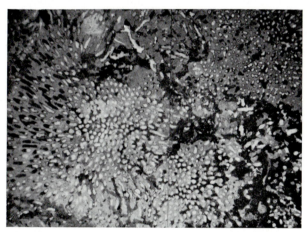

Fig. 19. Limestone-marble slabs of coral reef limestone. Sections across coral branches; white coral branches and red limestone matrix. Railroad station, Linz, Austria

h) *Chalk and chalky limestone* are soft, poorly lithified, white to grey evaporite carbonates of very fine grain size, high porosity, chalky structure and very low strength. Gradations are found towards a dense and sound limestone.

i) *Chert* may occur as nodules, flat lenticular bodies, or even as wavy continuous bands in carbonate rock. The concentration of silica has probably taken place during lithification. Chert always stands out in dimension stone by its more quartzlike, conchoidal fracture, but also by its sometimes different luster and color. Organic concentrations often colors chert dark to black, which is then known as flint. The presence of chert in soft limestone can cause considerable trouble in drilling operations if soft rock-drill bits suddenly and often unexpectedly hit the very hard chert. Flying silica dust from the chert requires maximum ventilation and wet drilling to avoid silicosis.

j) *Anhydrite*, the anhydrous calcium sulfate is usually associated with other evaporites, such as gypsum, some limestones, and salt rock. Anhydrite is relatively rare and thus plays only a subordinate role in the stone industry. Blue, translucent anhydrite was recently discovered in northwestern Indiana. Calcium sulfates are quite soft and water soluble; their use is therefore limited to interior decoration.

k) *Gypsum:* The hydrous calcium sulfate is, like anhydrite, associated with evaporite rock. Alabaster is very fine grained and translucent gypsum, often used as decorative and monumental material in Italy, Mexico, etc. The famous alabaster windows in Galla Placidia's Mausoleum in Ravenna, Italy, are of a beautiful yellow-to-orange color and the sole source of indoor light.

1.3. Metamorphic Rocks

The relocation of rocks through sediment accumulation and subsequent deep burial or crustal movements brings forth recrystallization towards more stable conditions. New minerals come into existence which chemically resemble the parent material but have "improved" towards a more stable form. Molecular

Table 3. *Origin and Mineral Composition of Metamorphic Rocks*

Low Metamorphic	High Metamorphic	Minerals	Parent Rock
Granite-gneiss	same	feldspar, quartz, hornblende	granite, arcosic sandstone, conglomerate
Hornblende-schist, greenstone	same	hornblende, mica, plagioclase	diorite, gabbro, shale, porphyry, basalt
Quartzite, fine	coarse	quartz, mica	quartz-sandstone, conglomerate
Marble, fine	coarse	calcite, (mica)	limestone, little clay
Marble-schist, fine	coarse	calcite, quartz, mica	lime-marl, clay-marl
Dolomitic marble fine	coarse	dolomite	dolostone, dolomitic limestone
Slate, calcareous	schist	mica, quartz, calcite	clay-marl
Slate	schist	mica, quartz	clay, shale

migration from one layer to another and beyond may take place during more prolonged exposure to metamorphic processes, leading to the formation of new minerals. The concentration of minerals along bands causes gneissic structure and foliation. High-temperature minerals are stable and may not lose their identity; the original massive, unoriented mineral arrangement, however, changes to a distinctly gneissic structure. The degree of metamorphic intensity influences the size of the individual mineral grains: low metamorphic zones lead to the formation of slates fine-grained marbles, and granite-gneisses, whereas higher metamorphic processes form schists and coarse-grained marbles. Table 3 presents common metamorphic rocks, their mineral composition and their possible parent rock.

Not all metamorphic rocks are of commercial stone quality: schists rich in mica split unevenly and are thus unfit; only certain gneisses, greenstones, marbles, quartzites and slates can be utilized.

Textures of Metamorphic Rocks

All mineral grains are tightly interlocked in metamorphic rocks; a serrated texture is common (Figure 10). The textures are limited to the degree of crystallinity as follows:

a) Microcrystalline or slaty texture is composed of submicroscopic mica flakes and some quartz, also of some other minerals which may arange in a parallel fashion.

b) Granular or granoblastic texture is composed of minerals of equal grain size, such as are found in marbles, quartzites and many gneisses.

c) Porphyroblastic texture is equivalent to the igneous prophyritic texture; the smaller grains literally wrap around the larger porphyroblasts.

Structures of Metamorphic Rocks

Metamorphic structures are dependent on the structures of the parent rocks, igneous or sedimentary. They reflect also the character and the degree of metamorphism. Metamorphic processes are gradational; so are the structures which they reflect, from igneous or sedimentary towards metamorphic.

a) *Massive:* granular rock with a massive appearance and without a visual orientation of the grains, such as marble, quartzite and some serpentines.

b) *Banded:* nearly parallel bands of minerals of different texture and color; characteristic for gneisses.

c) *Lineation:* the streaking on a foliated surface along which some movement has taken place. Hard minerals enclosed in a generally softer matrix cause the streaking along planes of schistosity or foliation under at least semi-plastic conditions.

d) *Foliation:* the splitting of the rock into thin sheets by the presence of tabular or prismatic minerals. Foliated rocks are also often lineated.

e) *Platy cleavage:* even foliation due to the parallel arrangement of submicroscopic sized mica. Cleavage can be irregular and incomplete because of interruption by small nodes and pits of quartz or other hard minerals. Imperfect cleavage also may result in a spiderweb-like plumose pattern on cleavage planes of split slate, discussed in the chapter on Natural Deformation of Rock and Stone (Figure 47 b). Slate surfaces marked with such features add character if used on walks or patios or as surface veneer on walls. Pennsylvania slate and most Vermont slates fall into the category of smooth slates, whereas Virginia slate is irregular.

f) *Ribbon slates:* Some slates, despite their slaty cleavage, still show the original bedding of the sedimentary structure in lines of darker or different color which cut across the present cleavage planes. Ribbon slates are quite ornamental on floors if the ribbons contrast with different colors. On the other hand, ribbons are considered flaws when they form lines of either weakness or greater hardness (Fig. 20).

g) *Folding:* Some gneisses and marbles are marked with tight folds formed by plastic or semiplastic flow during metamorphism. Folds in metamorphic rocks are generally not lines of weakness and they give a very desirable ornamental accent to the stone (see Fig. 21, also 52 of chapter, Natural Deformation of Stone).

Fig. 20. Two different ribbon slates: Red New York slate surrounded by white quartz gravel on walk, and loose slab of black slate from Bangor, Pennsylvania (right). Ribbons on New York slate are colored black and deep green, bands on Bangor slate are deep black. Ribbons intersect the slaty cleavage at about 65° (see arrow) on New York slate

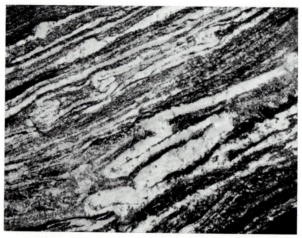

Fig. 21. Gneiss; light-colored bands of quartz and feldspar are flow-folded in plastic state like wet blankets. Folded metamorphic rocks are ornamental and if sound may be used as polished decorative stone or as ornamental blocks in landscaping. Glacial boulder, South Bend, Indiana

Mineral Composition of Metamorphic Rocks

The minerals of metamorphic rocks have the following origins:

a) Minerals inherited from igneous and sedimentary rocks, such as the feldspars, quartz, micas, hornblende, calcite and dolomite.

b) Minerals newly formed during the process of metamorphism, such as mica and sericite, chlorite, garnet and others.

c) Pigment minerals: hematite, magnetite, dark plagioclase and graphite. The chapter Color and Color Stability offers more information on rock colors.

Types of Metamorphic Rocks

a) *Gneiss* is a rock with a granitic look but with a more or less strongly developed parallel arrangement of the grains; gneissic structure should have a sound compressive strength across the grain (perpendicular to the long axes of the minerals) but generally a less favorable strength parallel to the grain. Layers of cleaving micas cause easy splitting of the stone along the mica bands. Gneisses are often classified as commercial granites.

b) *Schist* is similar to gneiss but has thinner banding (schistosity) than gneiss: the absence of the feldspars and quartz, and the increased presence of mica and hornblende facilitate rock splitting upon impact. The rock may have enough

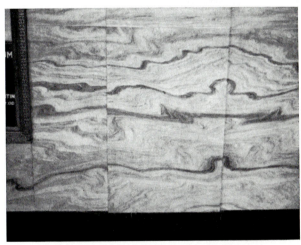

Fig. 22. Green marble, tightly flow-folded. Bottom strip is green serpentine. Shedd Aquarium, Chicago

strength to find application as rough slabs for wall veneers and flagstone, provided that the prismatic minerals prevail over the flaky micas. The stone industry often classifies schists as greenstone.

c) *Marble:* Metamorphic crystalline marble is the most frequently used metamorphic rock and it is the highest priced dimension and decorative stone. Composed of interlocking grains of fine or coarse calcite or dolomite, the color may be white, grey, pink, or green, often in streaks and bands (Fig. 22). The colors of true metamorphic marbles are "cold" as compared with the soft, warm pastel tones of sedimentary limestone-marbles. Very fine flakes of colored mica, graphite and some other minerals are the important carriers of the pigments; these pigments contrast the various shades of ferrous and ferric iron compounds as well as organic admixtures in sedimentary rocks. Pure white, fine-grained massive marble is often marketed as monumental, memorial or statuary stone. Coarser grained varieties, however, are less favorable, because the perfect cleavage of calcite in three directions may become an obstacle to precision sculpture work. White, sound marble has attracted famous sculptors through man's history; the Euboia marble of Greece and the Carrara marble of Italy (Figure 15) have served from ancient times through the Renaissance until today. In the USA, fine-grained Vermont marble resembles Carrara; coarse-grained white and pink marbles from Georgia and Alabama are often seen as exterior veneer stone.

d) *Serpentine*, often called serpentine-marble, is essentially composed of the mineral serpentine, a magnesium silicate; the stone is frequently used for its beautiful green color and texture. White veinlets of calcite and magnesite ($MgCO_3$) enhance the character of the stone, but the presence of magnesite may become a

Fig. 23. Cleavage piece of calcite compared with cleavage lines in calcite crystal. Calcite cleavage in coarse-grained marble can cause oriented chipping when worked by sculptors

serious problem in sulfate polluted air as the $MgCO_3$ converts readily to the very soluble $MgSO_4$. Great durability of the mineral serpentine permits both interior and exterior application. Despite its low hardness, serpentine should not be called a marble, but rather a greenstone. Some serpentines are translucent; the beautiful green color and the low hardness are inviting to sculptors. The names "Verde Antique", etc., remind us that the use of serpentine for ornamental purposes goes back to ancient times.

e) *Greenstones* are strongly metamorphosed basic igneous rocks, generally of green color resulting from the presence of green hornblende and chlorite as well as some other related minerals. Greenstone may be used like any other massive stone if it is not foliated.

f) *Quartzite* is a recrystallized quartz rock with interlocking grains of quartz, occasionally accompanied by some flakes of mica and grains of calcite. The rock is so hard that quarrying and milling is too costly to be competitive with other stone. Silica-cemented sedimentary quartz sandstones may also be called quartzites. The distinction is only minor, as both rock types are composed almost entirely of quartz and are of equal hardness and strength. The color of quartzite is generally white or light yellow, but it may also be red if iron oxides are present and greenish from the presence of mica. Free-silica dust in quartzite quarries should be considered a health hazzard which may cause silicosis of the lungs.

Table 4. *Major Rock Types Used in the Stone Industry*

Igneous Rocks:

	Intrusive	Intermediate	Extrusive
Light colored (acid)	granite		felsite
	syenite		felsite
Dark colored (basic)	diorite	porphyry	andesite
	gabbro		basalt

Sedimentary Rocks:

	Coarse	Medium	Fine
Clastic:	conglomerate, breccia	sandstone	shale
Organic:	shell limestone	crystalline and	limestone
	shell dolomite	shell limestone	dolomite
Evaporite:		crystalline	carbonates
		carbonates, gypsum	gypsum

Metamorphic Rocks:

	Coarse	Medium	Fine
From igneous rocks:	gneiss, schist	gneiss, schist	
From sedimentary rocks:	gneiss, schist	gneiss, schist	
	marble	marble	marble
		quartzite	quartzite
			serpentine

g) *Slates* are microgranular metamorphic rocks which were exposed to low-grade metamorphism. The slaty cleavage is generally of variable thickness. Great strength and durability are typical of good quality slates. Incompletely metamorphosed slates, however, show both the slaty cleavage and the original bedding planes which are preserved as "ribbons" (see Figure 20). The bedding planes may be both lines of weakness and areas of greater hardness compared with the slate matrix. Where exposed to heavy foot wear the ribbons may form a strong relief. Some slates cleave with feather-like markings which resemble spider webs on the surface. Irregular and incomplete cleavage caused by the presence of nodes and small pits on the cleavage surfaces, add character if used for walks or external surfaces.

The three major rock groups are summarized in Table 4. Common rock types which play a major role in the stone industry are cited for each group and subgroup.

References

1. BAYLY, B., 1968: Introduction to petrology. Prentice Hall, Inc., 371 p.
2. EICHER, D. L., 1968: Geologic Time. Prentice Hall, Inc., 149 p.
3. FOLK, R. L., 1957: Petrology of sedimentary rocks. University of Texas, Hemphills, Austin, Texas.
4. GRIFFITHS, J. C., 1967: Scientific methods in analysis of sediments. McGraw Hill, Inc., 508 p.
5. PARK, W. C., and E. H. SCHOT, 1968: Stylolites: Their nature and origin. Journal Sedimentary Petrology **38** (**1**), 175—191.
6. PETTIJOHN, F. J., 1957: Sedimentary rocks. Harper and Row, Inc., 2nd ed., 718 p.
7. WEYL, P. K., 1959: Pressure solution and the force of crystallization — a phenomenological theory. Journal Geophysical Research **64**, 2001—2025.

2. Properties of Stone

Stone is a heterogeneous substance characterized wide ranges of mineral composition, texture, and structure. Consequently, the physical and chemical properties, i. e., the durability, are extremely variable. The suitability of a stone for a given purpose should be based upon various properties that may be readily tested in the laboratory. Some properties are discussed and, where available, the ASTM (American Society for Testing and Materials) testing methods are included. ASTM specifications are given in Appendix B. At the end of this chapter, Table 6 summarizes technical rock properties.

A classification of the capillary system of rocks precedes such technical properties because the mean pore size determines the durability and moisture movement in stone. The sizes range from those of large vugs to those in apparently dense rock substance. A mean pore size is often difficult to establish in some stones, especially in dolomites with solution vugs which interrupt the fine pore system. After CHOQUETTE and PRAY (1970), megapores have a channel diameter of 256—0.062 mm., macrocapillaries (Vos, 1969) 0.062—0.0001 mm., and microcapillaries (Vos) 0.0001 mm. and smaller. An ideal dense rock substance does not exist in nature. Each pore size and pore shape reflects the origin of the rock.

2.1. Rock Pores

a) *Igneous rocks:* The capillary system permits some transmission of moisture despite very low water sorption, rarely in excess of 0.5%. Microcapillaries should be sought along mineral contacts and along cleavage planes of feldspars and ferromagnesian silicates. Micro-crack porosity in igneous rocks was described by NUR and SIMMONS (1970) in quartz-containing igneous rocks. During cooling from the melt, quartz contracts a total of about $4\frac{1}{2}\%$ or $3\frac{1}{2}\%$ from 573° C to atmospheric temperature. This value compares with only 2% or less for other common rock-forming minerals which accompany the quartz. This anomalous behaviour of the quartz tends to form micro-cracks around the quartz grains and across. The relatively high porosity of granites and their reduced strength should thus not surprise, despite the very high strength of quartz if uncracked.

b) *Sedimentary rocks:* The porosity and pore space distribution is subject to great variation based on the heterogeneity of sediments. Table 5 compares the kind and source of pores in sedimentary rocks after CHOQUETTE and PRAY (1970).

c) *Metamorphic rocks:* The parallel arrangement of flaky and prismatic minerals results in easy travel routes along mica flakes. Massive rocks such as some sound gneisses and marbles behave like igneous rocks.

Table 5. *Porosity of Sandstones and Carbonates*

Lithology	Sandstone	Carbonates
Original porosity		
uncemented	25—40%	40—70%
after lithification	15—30%	5—15%
Type of rock porosity	all interparticle	varied by post-depositional modification
Pore sizes, general	related to particle size, sorting	little related to particle size or sorting
Pore shape	by particle shape as negatives	varies greatly
Uniformity of size, shape	fairly uniform	variable, heterogeneous
Influence by lithification	minor: compaction, cementation	major: creates, obliterates, modifies
Semiquantitative evaluation	easy	variable, often impossible
Porosity-permeability relationship	consistent: depends on sorting, particle size	greatly varies: vugs and channels are unpredictable

From: CHOQUETTE and PRAY (1970).

2.2. Porosity

is the volume of pore space in the rock to its volume in per cent. Definitions of porosity for concrete by WALKER et al. (1969) also appear to apply to stone:

1. Total effective porosity, is the total pore volume determined by the mercury intrusion into the pore volume (V_v) at up to 1000 atm., measured in cm.^{3}. If n is the total effective porosity, V_b the bulk dry volume of the sample in cm.^{3}, then the equation follows, $n = \dfrac{V_v}{V_b} \times 100$.

2. Average porosity, is meant to comprise certain diameters, as the sum of the porosity between these diameters.

3. Pore modulus, is the cumulative sum of the penetration of fluid per gram of rock, corresponding to predetermined diameters expressed in cm.^{3}/gram.

4. Critical water content; no water transport is possible in capillary systems below this point.

The porosity number p is calculated as $\dfrac{B - A}{V} \times 100$, whereby A is the weight of the dried specimen, B the weight of the soaked and air-dried specimen, V the total volume of the sample. The porosity increases rapidly in clastic sediments with decreasing grain size. Clays may have a porosity well over 50% according to Fig. 24, after BAYLY (1968). The water in micropores is not readily available and cannot migrate. Porosity and bulk density ranges for some common rocks are given in Table 6.

Table 6. *Porosity and Bulk Density of Some Common Rocks*

Rock	Bulk Density (gm/cm.³)	Porosity (%)
Granite	2.6—2.7	0.5—1.5
Gabbro	3.0—3.1	0.1—0.2
Rhyolite (felsite)	2.4—2.6	4.0—6.0
Andesite (felsite)	2.2—2.3	10.0—15.0
Basalt	2.8—2.9	0.1—1.0
Sandstone	2.0—2.6	5.0—25.0
Shale	2.0—2.4	10.0—30.0
Limestone	2.2—2.6	5.0—20.0
Dolomite	2.5—2.6	1.0—5.0
Gneiss	2.9—3.0	0.5—1.5
Marble	2.6—2.7	0.5—2.0
Quartzite	2.65	0.1—0.5
Slate	2.6—2.7	0.1—0.5

From: FARMER (1968).

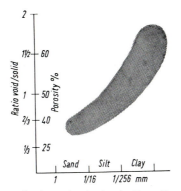

Fig. 24. Relationship of porosity to grain size in clastic sediments. From BAYLY (1968)

2.3. Water Sorption

is a general term which includes adsorption and absorption. Adsorption, according to WEBSTER, is the adhesion of molecules of gasses, or of ions or molecules in solutions, to the surfaces of solid bodies with which they are in contact. Absorption is the taking up, assimilation, or incorporation, as the absorption of gasses in liquids, as distinguished from "adsorption". The word absorption is frequently used to include adsorption.

The ASTM Standards (ASTM C-97-47, 1958) for the water sorption test for natural building stone suggests a least three smooth-surfaced specimens of 2 to 3″ size. After drying at 105° C for 24 hours and weighing, the specimens are weighed again after soaking in distilled water at 20° C for 48 hours. The percentage of absorption by weight equals $\dfrac{B-A}{A} \times 100$; A is the weight of the dried specimen and B the weight of the specimen after immersion in water.

2.4. Bulk Specific Gravity

or apparent specific gravity, is the ratio of the mass to that of an equal volume of
water at a specified temperature (ASTM C-97-47, 1958). The bulk specific gravity
is calculated as $\dfrac{A}{A-B}$; A is the weight of the dried specimen, B the weight soaked
and suspended in water. The bulk specific gravity multiplied by 62.4 yields the

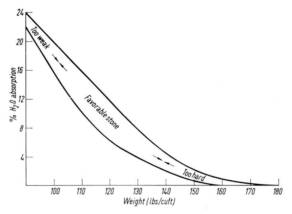

Fig. 25. Workability of stone, depending on the unit weight and water sorption. After Moen
(1967)

unit weight per cubic foot of dry stone. The stone industry measures the bulk
rock density as the dry weight in pounds per cubic foot. The test samples should
be similar to samples used in the absorption test. Moen (1967) approximates the
usefulness of commercial stone for external purposes on the basis of the bulk
specific gravity and the water sorption in Figure 25. Favorable stone ranges between 3 and 17% water content or a dry weight of 105—140 lb./ft.³.

2.5. Hardness

The hardness of a mineral or rock is its resistance to permanent deformation
and therefore an important factor for evaluating the workability of a stone in
quarry and mill and its resistance to mechanical wear. The hardness is very
consistent on fresh mineral surfaces and can be tested readily both in the field and
in the laboratory. We distinguish basically between the scratch hardness, the
indentaticn hardness, the abrasion hardness, the rebound hardness and the impact
hardness. The great complexity of rocks does not permit a close correlation of the
various strength and hardness parameters. Drilling operations deal with another
kind of hardness value than mechanical foot wear, or repeated impact by percussion. The most important types of rock (and mineral) hardness are discussed in
the following:

1. **Scratch hardness** is the ability of one solid to be scratched by another harder solid and is a complicated function of the elastic, plastic and frictional properties of a mineral surface. The scale was set up by Friedrich von Mohs in 1822 as an arbitrary scale based on 10 common minerals as standards: talc, as the softest mineral, was given $H=1$; diamond, the hardest mineral, $H=10$. The hardness intervals are wider spread at higher values if compared with a modern absolute scale such as the Vickers hardness. TAYLOR (1949) compares the Mohs hardness numbers with the Vickers Hardness. Frequently a revised and quantified Mohs hardness scale is used whereby the intervals from $H=1$ to $H=10$ are equal. These adjusted hardness numbers m are plotted in Figure 26 against the Shore rebound hardness, making possible decimal Mohs values.

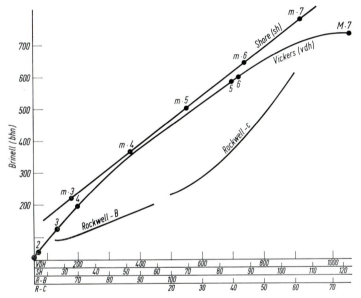

Fig. 26. Correlation of different hardness scales for minerals and metals.
Legend for abbreviations: M = original Mohs hardness, m = quantified Mohs hardness, bhn = Brinell hardness number, vdh = Vickers diamond hardness, sh = shore hardness, $R\text{-}B$ = Rockwell-B, $R\text{-}C$ = Rockwell-C

The scratch hardness of rocks is the oldest and simplest method to test rocks in the field. PROCTOR (1970) estimates the boring rate of tunneling machines using hardness ranges of common rocks, in Figure 27:

Granite and gneiss, with a range of 6 to 7, appear to be controlled by the presence of quartz.

Basalt and felsite, with a range of 5 to $6\frac{1}{2}$ appear to be influenced by the feldspar and hornblende.

Shale, ranging 2 to 3, is too soft for the stone industry.

Sandstone has the greatest range, from 2 to 7. The hardness of the cementing mineral as well as the degree of cementation determines the sandstone hardness.

Limestone and marble, except for chalky varieties, have hardnesses determined by the hardness of the mineral calcite. In dolomitic limestones, mixtures of the minerals calcite and dolomite hardness, reflects the proportion of the constituent minerals.

Slate, a very dense rock mostly composed of quartz and mica, ranges from hardness 3 to 5, depending on the mineral proportions and the degree of metamorphism.

Quartzite, both sedimentary and metamorphic quartzite, is the hardest rock with $H = 7$. A quartzose rock softer than $H = 7$ should not be called a quartzite.

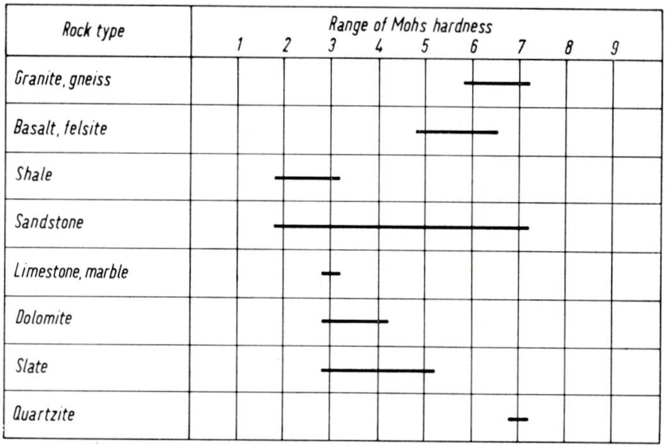

Rock type	Range of Mohs hardness
	1 2 3 4 5 6 7 8 9
Granite, gneiss	6—7
Basalt, felsite	5—6
Shale	2½—3
Sandstone	3—7
Limestone, marble	3
Dolomite	3½
Slate	3—5
Quartzite	7

Fig. 27. Rock hardness guide based on Mohs hardness. After PROCTOR (1970)

2. **Indentation hardness** is the permanent indentation of the mineral surface by a sphere, a cone, or a pyramidal indenter. The hardness is then determined by the load and the size of the indentation. Data of rock mechanics now often include indentation hardness and rebound hardness values. Instruments and techniques for the Brinell hardness, Vickers, and Knoop hardness were developed by the metallurgists.

a) *Brinell hardness* testing has been in use for testing metallic and non-metallic materials. A hard spherical indenter, a steel ball of a given diameter D, presses on a smooth mineral surface under a load W, in kg.; the mean chordal diameter of the resultant indentation measured gives d, in mm. The Brinell hardness number (BHN) is calculated,

$$\text{BHN} = \frac{W}{D(D-(D^2-d^2))}, \text{ in kg./mm.}^2.$$

b) The *Vickers Diamond Hardness* (VDH) is similar to the Brinell hardness. A pyramidal diamond indenter is pressed into the mineral surface under a load of W, in kg. The indenter encloses an angle of 136° between opposite faces, and 146° between opposite edges. The pyramidal area of the indentation is greater than the projected area of the indentation by the ratio 1 : 0.9272. Therefore the Vickers hardness is calculated:

$$\mathrm{VHD} = \frac{0.9272 \times W}{\text{projected area of indentation}},$$

which equals $1.8544 \ W/d^2$, in kg./mm.^{2}. The Vickers hardness is very similar to the Brinell hardness at low hardness numbers, but diverges at higher values where the BHN is higher. Indentation hardness values are often used in rock mechanics where they are compared with the grain size (Fig. 28), abrasion hardness for brittleness (Fig. 29), and percussion drilling hardness (Fig. 30). According to BRACE (1961) the indentation hardness increases with decreasing grain size. Basalt, quartzite, dolomite and limestone were investigated by BRACE. The plots, linear on a double log scale, were separate for each rock type. Numerous measurements had to be performed for meaningful averages.

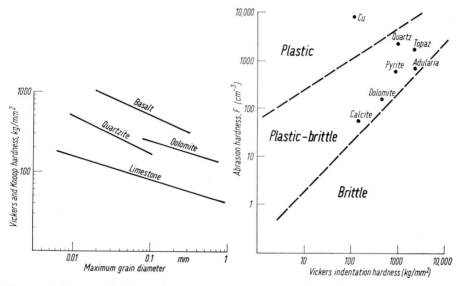

Fig. 28. Indentation hardness after VICKERS and KNOOP related to the maximum grain diameter of different rocks. After BRACE (1961)

Fig. 29. Brittleness of minerals as a function of the indentation and abrasion hardness. Simplified from ENGELHARDT and HAUSSUEHL (1965)

c) The *Knoop hardness*, H_k, is similar to the Vickers hardness in that a pyramidal diamond indenter is applied to the flat mineral surface. The indentation formed has the shape of a rhombus, with the conical angles of 172° 30′ and 130° (Fig. 31). The longer diagonal length does not change due to elastic recovery, and so it is used as the basis for the hardness measurement; the following equation applies:

$H_k = W/\text{area}$, executed at 0.2 to 4 kg. loads. The Knoop hardness number is very similar to the VDH.

d) The *Rockwell hardness* (R-B and the R-C test) are based on depth of penetration with a pre-load of 10 kg. which is first applied to the surface and is retained for the main test when the depth of indentation for a further 150 kg. is recorded on a dial gage. A spherical indenter is used for softer minerals (R-B test), a conical

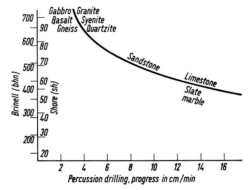

Fig. 30. Relationship of percussion drilling, as progress in cm/min, to Shore hardness (Brinell Hardness). The curve is an approximation. After MUELLER (1963) from KAHLER, modified by author

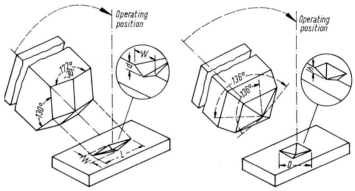

Fig. 31. Indentation hardness testers: Left: Knoop indenter. Right: Vickers pyramid Indenter. From LYSAGHT (1960)

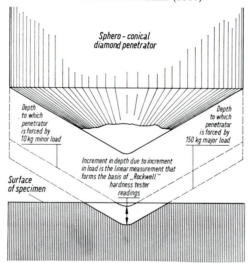

Fig. 32. Principle of indentation hardness testing after Rockwell. Courtesy of Wilson Instrument Division ACCO, 230 Park Ave., New York

indenter with a hemispherical tip for harder minerals (R-C test). The principles of penetration are shown in Fig. 32. Rockwell hardness tests are easiest to perform with good results for minerals and rocks apparently because pre-loading of the specimen eliminates errors through elastic recovery.

e) *Abrasion hardness* (abrasive strength) is the resistance of a mineral or rock to abrasive wear. The abrasion hardness for minerals was first developed almost to perfection by ROSIWAL (TERTSCH, 1949) in the late 1890's. A given quantity of abrasive is used without renewal. The abrasion time was 8 minutes because otherwise, without adding fresh abrasive, the abrasion rate slowed down too much to be practical. Due to great inconsistencies this abrasive hardness is little used today. ENGELHARDT and HAUSSUEHL (1965) modified the abrasion hardness test by using 24 r.p.m., 100 μ corundum powder and 100 g. total load on the specimen. The liquid selected could change the values greatly. In contrast, the abrasion hardness of rock specimens is used for the suitability of a rock under foot wear. More important than the abrasion hardness of the individual mineral fragments in rock is the resistance of the mineral bond and bonding agent to tearing. So, for instance, a weakly cemented quartz sandstone will record very low abrasion hardness despite the excellent abrasion hardness of the constituent quartz. The abrasion hardness value (ASTM C 241-51, 1958) H_a is the reciprocal of the volume of material abraded multiplied by ten. The superimposed weight of the specimen is 2000 g. plus the weight of the specimen.

$$H_a = \frac{10\,G\,(2000 + W_s)}{2000\,W_a};$$

where G is the bulk specific gravity of the sample, W_s the average weight of the specimen, and W_a the loss of weight during the grinding operation. The abrasion tests were developed by KESSLER from the old French Dorry abrasive resistance (H_d). Dense rock, such as finegrained limestone, records higher values than coarse grained varieties, because the grains loosen more easily along the larger interface area. Quartz sandstone with a weaker calcitic cement records H_a values closer to calcite than to quartz, because the much harder quartz grains are readily twisted from the cement. Thus the true abrasion hardness does not depend on the hardest mineral present but rather on the average mineral hardness and the strength of the interface bond.

f) *Brittleness* of minerals and rocks. Upon impact, a mineral may show a brittle or a plastic behavior. Calcite and quartz are considered brittle, copper plastic, malleable or ductile. ENGELHARDT and HAUSSUEHL (1965) found the brittleness of a mineral to be dependent on the Vickers indentation hardness and abrasion hardness. Fig. 29 presents this relationship on a double log scale. The plot distinguishes clearly three zones, i.e., brittle, plastic-brittle, and plastic. A few minerals are plotted into the graph. The presentation of the mineral brittleness gives a lead concerning the general brittleness of a rock. The rock brittleness is a very complex hardness parameter which resembles the mineral brittleness: it is composed of the abrasion hardness and the indentation hardness. Boundary surface contacts of the individual minerals, mineral hardness and the bond or cement influence the total brittleness. The brittleness of rocks is very difficult to measure unless a rock became "homogenized" by metamorphism.

g) *Shore scleroscope* hardness is a rebound hardness based on the rebound of a steel ball or of a diamond-pointed hammer which is dropped vertically into the test surface. The rebound is measured on an arbitrary scale of 120 divisions. The scleroscope hardness is a measure of the elastic properties of a rock or mineral; crushing decreases the rebound energy and thus the rebound height by an amount equal to the crushing energy and the energy absorbed otherwise by the rock surface and the instrument. The Shore hardness tester is a relatively inexpensive and compact instrument; its simplicity of operation permits many readings in a short time both in the laboratory and in the field. Shore hardness values are correlated with the Brinell, Rockwell, and Vickers hardness in Fig. 26; the values are approximations. The Shore hardness is compared with the compressive strength of rocks in Fig. 33, and with the rate of percussion drilling in Fig. 30. Differences in grain boundaries separate the relationship of granite, sandstone, and limestone Shore hardness to the compressive strength.

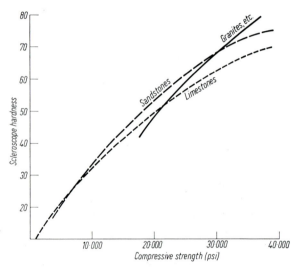

Fig. 33. Shore scleroscope hardness vs. compressive strenght. Different rock textures are responsible for the separation of granites, sandstones, and limestones. Compiled from data of OBERT *et al.* (1946), and WINDES (1949), and WINDES (1950)

h) *Schmidt hardness* is also a rebound hardness similar to the Shore hardness but it is obtained with a concrete Schmidt test hammer which is designed to estimate the strength of concrete in place on a structure. The Schmidt L-type hammer impact energy of 0.54 ft.-lbs. was chosen because softer rock specimens broke upon impact at higher energies. Good correlation was obtained with the Shore hardness as both methods operate on the same principle. The relationship of the Schmidt hardness with the uniaxial compressive strength is semi-logarithmic; Fig. 34 simplifies DEERE's (1968) data. The graph separates rocks of different dry unit weight.

i) *Impact toughness* is the resistance of rock to sudden impact; curb stone and aggregate stone in concrete in heavy traffic may be readily exposed to such impact. The rock impact toughness (ASTM D 3—18, 1942) is produced by a plunger which in turn is hit by a 2-kg. drop weight on a specimen 25 mm. in dia-

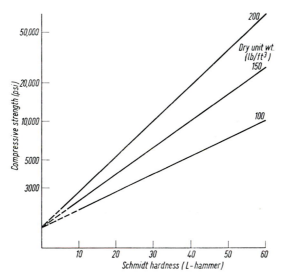

Fig. 34. Relationship between Schmidt Impact Hardness, uniaxial compressive strength, and the dry unit weight. Simplified from DEERE (1968)

meter and 25 mm. in height. The impact toughness is expressed by the minimum height at which the drop weight breaks the specimen, measured in inches per square inch. Fig. 35 compares the impact toughness of rock with the compres-

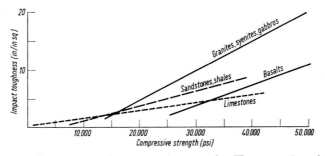

Fig. 35. Impact toughness versus the compressive strenght. The separation of the different rock groups is similar to the relationship of the scleroscope hardness versus the compressive strength. Data compiled from data of OBERT *et al.* (1946), and WINDES (1949), and WINDES (1950)

sive strength. At a given compressive strength granite has a higher impact toughness than sandstone or limestone. Differences in intergranular bonding appear to have an influence similar to that on the Schmidt and the Shore hardness.

j) *Drilling hardness,* or the rock drillability, has interested the stone industry as accurate cost estimates of drilling operations are sought. The drilling hardness of minerals was experimentally measured by PFAFF, brought to attention and criticized by TERTSCH (1949). A rotary diamond point drills a depression into the

Table 7. *Mineral Hardness Values, Compared with the Mohs Hardness Scale*

	Mohs	Vickers	Knoop	Abrasive	Drilling
Talc	1	47	—	0.003	—
Gypsum	2	60	46—54	1.25	8.3
Calcite	3	105—136	75—120	4.5	50
Fluorite	4	175—200	139—152	5.0	143
Orthoclase	6	714	560	37	4665
Quartz	7	1103—1260	666—902	120	7648
Topaz	8	1648	1250	194	28,867
Corundum	9	2085	1700—2200	1000	188,808
Diamond	10	—	8000	140,000	—

Compiled from: TAYLOR (1949), WINCHELL (1945), TERTSCH (1949).

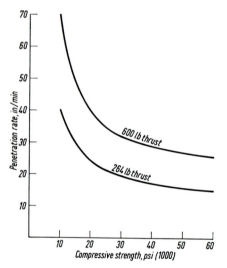

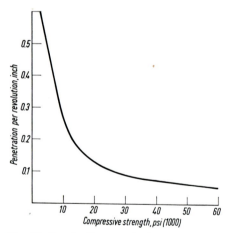

Fig. 36. Penetration rate in percussive drilling, given for experimental thrusts of 600 lb, and 264 lb, as a function of the compressive strength of rocks. From BRUCE (1970)

Fig. 37. Penetration rate in diamond drilling as a function of the compressive strength of rock. From PAONE and BRUCE (1963)

mineral surface at 6 to 7 revolutions per second. The drilling hardness consists of the number of revolutions necessary to cut a hole 10 μ deep. The drilling hardness is proportional to other hardness values as summarized on Table 7.

Percussion drilling and rotary drilling are the basic types of drilling in rock; important mechanical variables are involved such as rock hardness, impact tough-

ness, drill crown or drill bit sharpness and drillshaft pressure; a prediction of the
drilling speed is therefore very difficult. This uncertainty is still further enhanced
by the scarcity and great scatter of basic rock mechanics data. The drilling
advance, or the drilling volume per unit time, is approximately proportional to
the power or energy exerted and to the blow frequency.

Percussive drilling is the most common drilling in the stone industry. BRUCE
(1970) summarizes the available information, estimating the percussive drilling
rate with the equation, $R = f(E, 1/P)$, whereby R is the penetration rate in
inches per minute, E the work rate in in.-lb. per minute, and P the rock physical
property. The penetration rate R is plotted against the compressive strength in
Fig. 36 for thrusts of 600 lbs. and 264 lbs. The rate R is inversely proportional to
the rock strength. The relationship of the drilling progress against the Shore
impact hardness is given in Fig. 30; the data should be used with caution, however.

Diamond drilling is used for the recovery of drill cores to study the physical and
chemical rock properties. PAONE and BRUCE (1963) estimate the drill bit penetration in inches per revolution and the relationship to the compressive strength
of the rock, in Fig. 37. The curve is based on a theoretical equation with only
little deviation from field measurements.

2.6. Compressive Strength

is the load per unit area under which a block fails by shear or splitting. The
compressive strength is a very important parameter in rock mechanics, both
unconfined (uniaxial) and confined (triaxial). Rock strength is unconfined near

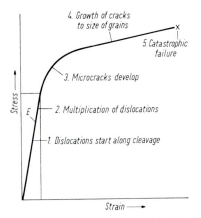

Fig. 38. Stress-strain curve in compression and tension of brittle fracture. E, the modulus of
elasticity, is the ratio of stress to strain. The progressive destruction of test specimens is
marked on the diagram. Modified from CONRAD and SUJATA (1960)

the earth's surface, also on buildings, whereas confined compressive strength
reflects rock strength in mountain-building processes, in deep rock foundations
and tunnels. The ASTM Standards (C 170-50,1958) recommend uniaxial strength

testing as follows: three or more specimens should be used not smaller than 2″, coarse grained rock not less than 2,5″ in size. Testing should be parallel to the bedding, either dried at 105° C for 24 hours, or wet after immersion in 20° C water for 48 hours. The compressive strength is then calculated as $C = \dfrac{W}{A}$; where C is the compressive strength of the specimen in psi, W is the load in pounds on the specimen at failure, A is the calculated area of the bearing surface in square inches. The rate of loading should not exceed 100 psi/sec.

Uniaxial compression as well as tension develop a basic and characteristic stress-strain diagram. In the first 2 stages of Fig. 38, dislocation along cleavage planes takes place. The modulus of elasticity E is calculated from this part of the

Term	Psi	kg/cm	Rock materials
Very weak	<1000	<70	
Weak	1000 – 3000	70 – 200	
Medium strong	3000 – 10,000	200 – 700	
Strong	10,000 – 20,000	700 – 1400	
Very strong	> 20,000	>1400	

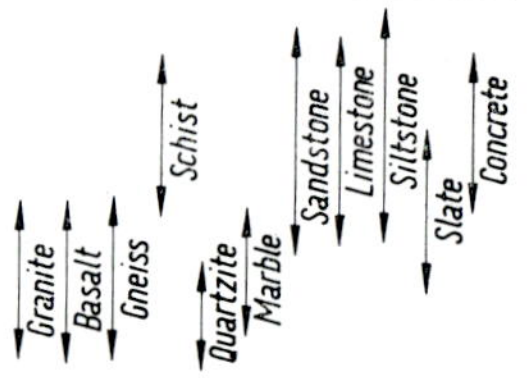

Fig. 39. Strength classification of rock materials on the basis of compressive strength. After HAWKES and MELLOR (1970)

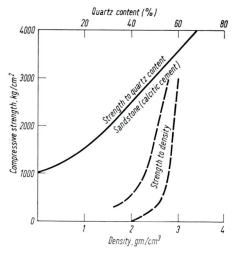

Fig. 40. Influence of the quartz content of sandstones and the density of rocks on the compressive strength. From FARMER (1968)

stress-strain curve (see Modulus of Elasticity of this chapter). When microcracks start to develop, the strain increases rapidly leading to the growth of cracks to the size of the grains. Catastrophic failure is the final stage of compression or tension.

The uniaxial compressive strength of rock can be used as the general index for rock strength. Fig. 39 presents a general strength classification after HAWKES and MELLOR (1970). In general, decreasing grain size increases the compressive strength. DREYER (1966) found a major increase of strength in rock salt with decreasing grain size in the grain size range 0—30 grains/cm.². Further decrease of the grain size from 30 to 100 grains/cm.² was minor. DREYER's observation can well be extrapolated to other homogenous rock types.

Bedded and foliated rocks record different rock strengths depending on the angle of bedding or foliation to the direction of stress; the maximum strength is usually obtained at 30°, the minimum strength between 0 and 15°.

Compressive strength is an important parameter in rock mechanics for comparing other rock strength values. Fig. 40 compares the compressive strength with both the rock density and the amount of quartz present in a sandstone with calcitic cement. The compressive strength is also related with the tensile strength (see chapter on tensile strength). Comparisons with the scleroscope hardness of impact toughness are limited as granite can only be compared with other igneous rocks, sandstone with other sandstones etc.

2.7. Tensile Strength

is the degree of coherence of the rock to resist the pulling force; this depends both on the strength of the mineral grains and on the cement, or the interface area from one mineral to the next. BRACE (1964) believes that brittle

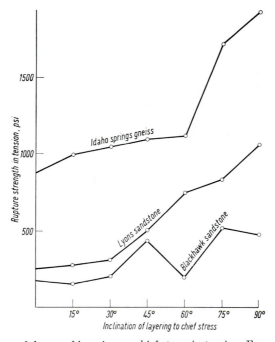

Fig. 41. Influence of degree of layering on chief stress in tension. From YOUASH (1969)

fracture apparently starts at the grain boundaries, which loosen and become partially detached as tension continues and fracture approaches. Thus tensile strength and grain size dependence of strength agree roughly with that predicted by GRIFFITH. HARDY and JAYARAMAN (1970) summarize different testing techniques and results of tensile strength of rocks and present an improved and consistent Hoop-Stress method: A hollow disc is exposed to an expanding rubber insert into the central hole of the doughnut shaped specimen; expansion within the rubber tube exposes the disc to tension. Tensile strength is a very important parameter for the estimate of resistence of rock to expanding salts and freezing water. The tensile strength was estimated by GRIFFITH (1924) to be one-seventh to one-eleventh the compressive strength. The improved Hardy test for tensile strength will establish more consistent data. YOUASH (1969) plots the relationship between the rupture strength as tension against the inclination of the test specimens to bedding and gneissic structure. As expected, maximum strength values were obtained when bedding or other planes of weakness were 90° to the vertical axis, or parallel to the direction of the tensional stresses, as shown on Fig. 41.

2.8. Modulus of Elasticity

Rock properties are governed by the reaction of the rock to the forces acting on it; forces induce a state of stress which results in deformation, i.e., a state of strain. The relationship between stress and strain is the basis for the Modulus of Elasticity, E. The E-value is graphically explained on the stress-strain diagram

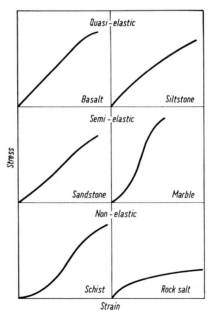

Fig. 42. Stress-strain curves for some common rock types. Modified from DEERE (1968) and FARMER (1968)

of Fig. 38. Strain of an ideally elastic substance recovers fully after removal of the stress. The linear stress-strain relationship is the modulus of elasticity expressed as $E = \sigma/\varepsilon$, where σ is the stress, and ε the rate of strain. The limit of elastic deformation is the strength in brittle material. In general fine-grained rocks are quasi-elastic, whereas coarser grained still cohesive rocks are semi-elastic. Coarse-grained rocks with high porosity are non-elastic. Stress-strain curves for some common rocks are sketched in Fig. 42 and E-values are given in Table 8.

Table 8. *E-Values of Some Important Rocks at Zero Load*

Rock	E (kg./cm.²), $\times 10^5$	Rock	E (kg./cm.²), $\times 10^5$
Granite	2—6	Microgranite	3—8
Syenite	6—8	Diorite	7—10
Gabbro	7—11	Basalt	6—10
Sandstone	0.3—8	Shale	1—3.5
Limestone	1—8	Dolomite	4—8.4

From: FARMER (1968).

2.9. Rock Creep

is the time-dependent deformation of a rock substance subjected to constant stress conditions. Creep is accelerated by well developed perfect cleavage of calcite in 3 directions. Marble is therefore very susceptible to creep and deformation possible on freely suspended marble slabs by their own weight. Buckling

Table 9. *E-Values and Creep Rates of Various States of Elasticity*

Degree of Elasticity	E-Value	Creep Rate (at 100 kg./cm.² stress, in 10 years)
Quasi-elastic	12	7.6×10^{-7}
	10	1.0×10^{-6}
	8	1.4×10^{-6}
Semi-elastic	6	2.1×10^{-6}
	4	4.0×10^{-6}
Non-elastic	2	1.1×10^{-5}
	0.5	8.9×10^{-5}

Adapted from: FARMER (1968).

by creep on marble slabs of New Orleans grave yards and other old cemeteries of the Old and New World are a common sight (see Fig. 53, Deformation of Rocks and Stones). The creep rates are influenced by the E-value of a rock; creep is therefore high in non-elastic rocks, highest in rock salt. Table 9 gives some ranges.

2.10. Modulus of Rupture

or flexural strength, is the resistance of a rock slab to bending or flexure (ASTM C 99-52, 1952). Wind stresses, snow load, and stacking loads of stone slabs in the mills may bring forth such stresses. The flexural test is executed by center-point loading of a knife edge which should be performed parallel and perpendicular to bedding or banding, the knife edge being as long as the width of the specimen.

The modulus of rupture is then calculated with the simple formula $R = \dfrac{3\,W\,l}{2\,b\,d}2;$

where R is the modulus of rupture in psi, W is the breaking load in pounds, l is the span length between supporting knife edges in inches. The required sample dimension of $4 \times 8 \times 2\frac{1}{4}''$, is tested either dry or wet.

2.11. Thermal Expansion of Minerals and Rocks

important where stone is exposed to rapid changes of temperatures, both in diurnal heat cycles and in fires. (See also the chapter on Fire Resistance of Minerals and Rocks.) Some damage to stone on buildings has been ascribed to differential thermal expansion. Quartz, for instance, can produce 545 kg./cm.^{2} expansion pressure if heated 40° C; this difference of temperature is not exceptional in desert or urban environments. HOCKMAN and KESSLER (1950) believe that some damage done to granites is possibly the result of heating-cooling cycles; damage was usually observed if heating and cooling was combined with moisture

Table 10. *Average Thermal Expansion at Low Temperatures for Limestone and Dolomite*

Temperature (° C)	Expansion by Volume ($10^{-6}/°$ C)	
	Limestone	Dolomite
−20 to +20	3.4	7.7
+20 to +80	8.1	9.9

From: VILLWOCK (1966). The K values of the American scale are conversions of data given by VILLWOCK in the metric system.

expansion. Calcites in marbles with crystallographic orientation expand parallel to the c-axis and at the same time contract perpendicular to it, even at temperatures below 100° C. (See Fire Resistance of Minerals and Rocks.) Limestones are only about half as expansive as dolomites at temperatures between −20 to +20°C, according to HARVEY (1967), Table 10. Areas of strong climatic contrasts in some moderate and subarctic climate have to be cautious of dolomites where differential expansion may take place even for concrete aggregates.

2.12. Thermal Conductivity

reflects the insulating capacity of stone, a property important for stone as building material. The guarded-hot-plate method (ASTM C 177-63) for testing flat slabs can be applied to stone though designed for building materials in general. The thermal conductivity can be calculated by the formula $K = \dfrac{q\,L}{A\,(t_1 - t_2)}$; where K is the thermal conductivity in Btu in. per hr./sq. ft./deg. Fahr, q is the rate of heat flow in Btu per hr., L is the thickness of specimen in inches measured along a path normal to isothermal surfaces, A is the area of isothermal surface in square feet, t_1 is the temperature of the hot surface in degrees F and t_2 the temperature of cold surface. Dense rocks have, in general, a higher thermal conductivity than porous rocks. Table 11 gives some K values for common rocks, as well as for air and water.

Table 11. *Thermal Conductivity Values of Some Common Rocks*

Rock	K (cal cm. per sec/cm.^{2}/° C)	or: K (Btu in. per hr/sqft/° F)
Granite	4—8	11.612—23.224
Basalt	3—7	8.709—20.321
Sandstone	3—8	8.709—23.224
Limestone	5—8	14.515—23.224
Gneiss	4—5	11.612—14.515
Marble	5—6	14.515—17.418
(air)	0.0057	16.55
(water)	0.0072	20.90

From: Villwock (1966). The K values of the American scale.

2.13. Light Transmission

Thin, translucent slabs of stone have been used as a substitute for glass windows in the Mediterranean area since ancient times. The alabaster windows of Galla Placidia's Mausoleum in Ravenna, the onyx windows of St. Paul's in Rome, and others have not only provided subdued interior lighting, spreading hues of white, buff, pink and light green, but also lend an atmosphere of distinction and warmth. Translucent colored marble panels are again fashionable, resembling stained glass windows with nature's own manifold designs.

Light transmission through a homogenous substance such as glass or fine-grained marble is calculated by the equation $I = I_0 e^{t\,x}$, where t is the light transmittance, I_0 the intensity of light entering the slab, I the intensity of the transmitted light, and x the thickness of the slab; the equation can be modified to $t = \dfrac{(\log I - \log I_0)}{x}$ which defines the slope of the line in per cent, with the light transmission on a log scale, the slab thickness on linear scale. The light transmission data for a few commercial marbles and for highly translucent Mexican Onyx are plotted in Fig. 43. The graphs permit extrapolation of the light transmission for any thickness of the marble slab.

The light transmission may also be evaluated on a qualitative basis, as transparent, translucent and opaque.

Transparent

A material is transparent if an object behind a thin slab of the material can be clearly identified, like glass, clear white mica, quartz crystals, some gypsum crystals, etc. Transparent rock is unknown.

Translucent

Light is readily transmitted through a thin sheet or an edge, but objects behind a slab cannot be identified. Many marbles and onyx-marbles are translucent.

Table 12. *Properties of Rocks in Common Use*

	Mohs H H_m	Shore H S_h	App. Spec. Grav.	Porosity %
Granites	5.80—6.60	85—100	2.54—2.66	0.4—2.36
Syenites	5.68—6.58	82—99	2.72—2.97	0.9—1.9
Gabbros, diorite, diabase	4.76—6.21	40—92	2.81—3.03	0.3—2.7
Basalt	4—6	50—92		
Limestone	2.79—4.84	10—60	1.79—2.92	0.26—3.60
Sandstone	2.40—6.1	20—70		
Gneiss	5.26—6.47	74—97	2.64—3.36	0.5—0.8
Quartzite	4.2—6.6	55—83	2.75	0.3
Marble	3.7—4.3	45—56	2.37—3.2	0.6—2.3
Slate		45—58	2.71—2.9	0.1—4.3
Quartz				
Orthoclase				
Albite				

Compilation from: CLARK and CANDLE (1961), WINDES (1949, 1950), BLAIR (1955, 1956).

Opaque

Light cannot pass through the material, even through thin edges. Most rocks are opaque.

The amount of transmitted light through crystalline calcite marbles depends on the optical orientation of calcite, micas and other minerals with relation to the slab. Scattering of light occurs along the crystal boundaries, by tiny included gas bubbles and by pigments. The light transmission appears to be higher perpendicular to the optical axis (into the rock banding) than parallel to it. Mexican Onyx is an exception, as it is more translucent perpendicular to the banding (WINKLER and SCHNEIDER, 1965). Grain size, pigment, inclusions and mean grain orientation are a major factor in the light transmission of marbles. A wet or polished marble surface disperses less light than a dry surface; polishing also increases the light transmission.

Table 12 summarizes the ranges of properties of stone varieties in common use.

Table 12. *(Continued)*

Comp. Strength 10^3 psi	Modulus Rupture 10^3 psi	Impact Toughness in/in^2	Abrasion Hardness H_a	Thermal Expansion 10^{-7}/° C
14—45	1.3—5.5	7—28	37—88	37—60
27—63	2.3—3.2	6.3—14		(37)
18—44	2—8	5.6—34		20—30
16—49	2—8	5—40		22—35
2—37	0.5—5.2	5—8.6	2—24	17—68
5—36	0.7—2.3	2—35	2—26	37—63
22—36	1.2—3.1	3.7—8.4		13—44
30—91	1.2—4.5	5—30		60
10—35	0.6—4	2—23	7—42	27—51
20—30	5—16		6—19	45—49
			180	
			53	
			81	

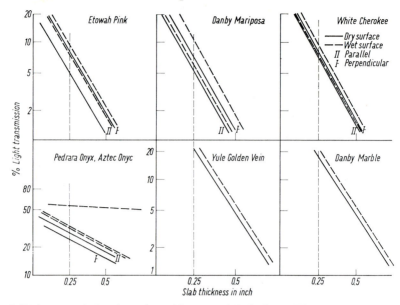

Fig. 43. Light transmission through marble versus slab thickness. The influence of the surface finish is indicated by differences in light transmission through a wet or polished marble surface and a dry, honed or finely ground surface. From WINKLER and SCHNEIDER (1965)

References

1. ASTM, 1969: Standards, part 10, 12, 14. American Society Testing and Materials, Philadelphia, Pa.
2. BAYLY, B., 1968: Introduction to petrology. Prentice Hall, Inc., 371 p.
3. BERRY, B. E., and B. MASON, 1959: Mineralogy. Freeman and Co., 630 p.
4. BRACE, W. F., 1961: Dependence of fracture strength of rocks on grain size. Proceedings of the Fourth Symposium Rock Mechanics, Pennsylvania State Univ., March 30—April 1, 1961, pp. 99—103.
5. BRACE, W. F., 1964: Brittle fracture of rocks. In: State of Stress in the Earth's Crust (W. R. JUDD, ed.); American Elsevier Publishing Co, pp. 111—180.
6. BRUCE, W. E., 1970: How to predict the penetration rate of percussion drills. Mining Engineering, November 1970, pp. 62—64.
7. CLARK, G. B, and R. D. CANDLE, 1961: Geologic structure stability and deep protection construction. Technical Documentary Report No. AFS WC-TDR-61-93. Air Force Special Weapons Center, Project No. 1080, Task No. 108001.
8. CONRAD, H., and H. L. SUJATA, 1960: Dislocation theory applied to structural design problems in ceramics. National Academy Science, Materials Advisory Board, Dec. 1960.
9. DEERE, D. U., 1968: Rock mechanics, geological considerations. In: Rock Mechanics in Engineering Practice (K. G. STAGG and O. C. ZIENKIEWICZ, eds.), London-New York: Wiley, pp. 1—53.
10. DREYER, W., 1966: Quantitative Untersuchungen über die Festigkeit einfach strukturierter Gesteinsarten in Korrelation zu den Gefügeparametern und dem Mineralgehalt der Akzessorien. Proceedings 1st Congress of the International Society of Rock Mechanics, Lisbon, **1**, 133—142.
11. ENGELHARDT, W. V., und S. HAUSSUEHL, 1965: Festigkeit und Härte von Kristallen. Fortschr. Mineralog., **42** (5).

12. FARMER, I. W., 1968: Engineering properties of rocks. London: E. and F. N. Spon Ltd., 180 p.
13. GRIFFITH, A. A., 1924: Theory of rupture. Proceedings 1st International Congress of Applied Mechanics, Delft, pp. 55—63.
14. HARDY, H. R., and N. I. JAYARAMAN, 1970: An investigation of methods for the determination of the tensile strength of rock. Proceedings 2nd Congress International Society of Rock Mechanics, Belgrade, Yugoslavia, Sept. 1970, 3, paper 5/12.
15. HARVEY, R. D., 1967: Thermal expansion of certain Illinois limestones and dolomites. Illinois State Geological Survey Circular 415, 33 p.
16. HAWKES, I., and M. MELLOR, 1970: Uniaxial testing in rock mechanics laboratories. Engineering Geology, 4 (3), 177—285.
17. HOCKMAN, A., and D. W. KESSLER, 1950: Thermal and moisture expansion studies of some domestic granites. U.S. National Bureau of Standards, Research Paper 2087, 44, 395—410.
18. HOEK, E., 1968: Brittle fracture of rock. In: Rock Mechanics in Engineering Practice (K. G. STAGG and O. C. ZIENKIEWICZ, eds.), London-New York: Wiley, pp. 99—124.
19. LYSAGHT, V. E., 1960: The how and why of microhardness testing. Metal Progress, 1960, August.
20. MOEN, W. S., 1967: Building stone of Washington. Washington Department of Conservation, Division of Mines and Geology, Bull. 55, 85 p.
21. MUELLER, L., 1963: Der Felsbau. Stuttgart: Ferd. Enke, 624 p.
22. NUR, A., and G. SIMMONS, 1970: The origin of small cracks in igneous rocks. International Journal Rock Mechanics and Mining Science, 7, 307—314.
23. OBERT, L., S. L. WINDES, and W. I. DUVALL, 1946: Standardized tests for determining the physical properties of mine rock. U.S. Bureau of Mines, Report of Investigation 3891, 67 p.
24. PAONE, J., and W. E. BRUCE, 1963: Drillability studies, diamond drilling. U.S. Bureau of Mines, Report of Investigation 6324, 32 p.
25. PROCTOR, R. J., 1970: Performance of tunnel boring machines. Bulletin Assoc. Engineering Geologists, VI. (2), 105—117.
26. VILLWOCK, R., 1966: Industriegesteinskunde. Offenbach/Main, Germany: Stein-Verlag, 279 p.
27. WALKER, R. D., H. C. PENCE, W. H. HAZLETT, and W. J. ONG, 1969: One-cycle slow freeze test for evaluation of aggregate performance in frosen concrete. Natl. Cooperative Highway Research Program Report, 65, 21 p.
28. WINDES, S. L., 1949: Physical properties of mine rock, part I, U.S. Bureau of Mines, Report of Investigation 4459, 79 p.
29. WINDES, S. L., 1950: Physical properties of mine rock, part II, U.S. Bureau of Mines, Report of Investigation 4727, 37 p.
30. WINKLER, E. M., and G. J. SCHNEIDER, 1965: Light transmission through structural marble. American Institute of Architects, Journal, March 1965, pp. 67—68.
31. YOUASH, Y., 1969: Tension tests on layered rocks. Geological Society America, Bulletin, 80 (2), 303—306.

Important Additional References

1. AGI, 1957, 1960: Glossary of geology and related sciences. The Am. Geol. Inst., Washington, D. C., 365 p., plus Supplement (1960), 71 p.
2. Anonymous, 1962: Marble Engineering Handbook. Marble Institute of America, 119 p., Washington, D. C.
3. Anonymous, 1968: A dictionary of mining, minerals, and related terms. Div. of Public Documents, U.S. Govt. Printing Office, Washington, D. C.

4. BARTON, W. R., 1968: Dimension stone. U.S. Bur. of Mines, Circular 8391, 147 p.

5. BAYLY, B., 1968: Introduction to petrology. Prentice Hall, Inc., 371 p.

6. BLAIR, B. E., 1955: Physical properties of mine rock, part III. U.S. Bur. of Mines, Rept. of Investig. 4727, 37 p.

7. BLAIR, B. E., 1956: Physical properties of mine rock, part IV. U.S. Bur. Mines, Rept. of Investig. 5244, 69 p.

8. BOWLES, O., 1956: Granite as dimension stone. U.S. Bur. of Mines, Information Circ. 7753, 18 p.

9. BOWLES, O., 1956: Limestone and dolomite. U.S. Bur. Mines, Inf. Circ. 7738, 29 p.

10. BOWLES, O., and W. R. BARTON, 1963: Sandstone as dimension stone. U.S. Bur. of Mines, Inf. Circ. 8182, 30 p.

11. BOWLES, O., and R. L. WILLIAMS, 1963: Traprock. U.S. Bur. of Mines, Inf. Circ. 8184, 19 p.

12. COURRIER, L. W., 1960: Geologic appraisal of dimension stone deposits. U.S. Geol. Survey, Bull. 1109, 78 p.

13. Dictionary of geological terms, 1962: AGI, A Dolphin Reference Book, Garden City, N.Y.: Doubleday and Co., 545 p.

14. INGRAM, S. H., 1963: Roofing rock: quality and specifications. Calif. Div. Mines and Geology, Mineral Inf. Service, 16 (1), 1—7.

15. KESSLER, D. W., 1933: Wear resistance of natural stone flooring. Natl. Bur. Standards, Jour. of Research Paper 612, pp. 635—648.

16. KESSLER, D. W., and W. H. SLIGH, 1932: Physical properties and weathering characteristics of slate. Natl. Bur. Standards, Jour. of Research Paper 477, pp. 377—411.

17. KESSLER, D. W., H. INSLEY, and W. H. SLIGH, 1940: Physical, mineralogical and durability studies on the building and monumental granites of the United States. Natl. Bur. Standards Research Paper 1320, pp. 161—206.

18. MOOS, A. VON, und F. DEQUERVAIN, 1948: Technische Gesteinskunde, Basel: Verlag Birkhaeuser, 221 p.

19. NELSON, A., and K. D. NELSON, 1967: Dictionary of applied geology, mining and civil engineering. London W. C. 2: Georges Newnes Ltd.

3. Natural Deformation of Rock and Stone

Most rocks were exposed to stresses in the earth's crust in the geologic past. Such stresses cause cracking and faulting under near-surface brittle conditions, folding if more plastic, as at greater depth. The stresses are compressional, tensional or shear. The variety and combination of deformational features is almost unlimited and the basic concepts of faulting and folding are found in basic text-books of physical geology, to which the reader is referred. This book deals with only those simple rock structures which influence the stone appearance, the stone quality and quarry operations.

3.1. Brittle Rock Fracture

Rock fracture is most significant, because it can determine the size of the stone and mining safety.

Jointing

A joint is a fracture in rock, generally more or less vertical, along which no appreciable movement has occurred. Joints determine the minimum size of the stone which the quarry operator can recover. Joints may be aids to the economic feasibility of hard-rock quarrying. Rock joints are ascribed to tension, extension and

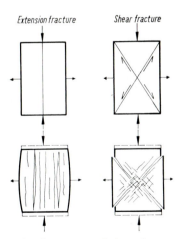

Fig. 44. Development of extension fractures and shear fractures. Extension fractures often show plumose markings on joint surfaces, whereas shear fractures may develop slickensides by minor movements

shear. Fig. 44 displays the basic fracture pattern during unconfined compression of a cube or cylinder of stone or concrete. The angles between the fractures enclose almost 90°. A great complexity of multidirectional jointing in rock masses can develop from higher order shear fractures. Joints generally reflect predominant regional structural trends. Hodgson (1961) classifies joints aptly as systematic, cross joints, and predominant bedding planes (Fig. 45). Non-systematic joints are usually curved and terminate at systematic joints.

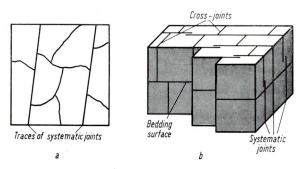

Fig. 45. Basic joint patterns
A. Planview of the relationship of systematic joints to non-systematic joints
B. Schematic block diagram showing relations between cross joints, systematic joints and bedding surfaces

From Hodgson (1961)

The joint magnitude, joint width and joint filling (with clay or crushed mineral substance) play very important roles in quarrying and quarry safety. Ornamental effects of colors and markings may occur along joints which have been filled and repaired with bright white secondary calcite or quartz. Mueller (1963) classifies large joints as an expression of geological disturbances; they may be simple joints, joint bundles, joint swarms, large joints, feather joints, and shear zones. Shear zones and feather joints indicate shear movement; such joints cut across crystals and are frequently slickensided (grooved, striated or polished) whereas tension fractures show clear, granular breaks sometimes filled with clay. Clay fillings can range from mere films to major zones.

Colors on Joint Surfaces

Weathering readily enters and penetrates along joints and discolors the surface with a wide range of colors of iron oxide and hydroxide. A joint surface colored or discolored in this way can contribute to a variety of colors. Interesting architectural effects of color can be obtained by properly mixing a variety of color shades on a wall (see color plate Fig. 68).

Fabrics on Joint Surfaces

Feather-like markings (plumes) are often found on smooth joint surfaces. The plumes diverge from a central axis and pass into a plumose (feathery) system of minor planes. Fig. 46 illustrates the basic concept after Hodgson (1961) and

ROBERTS (1961). Plumose patterns are not only visible on natural joint surfaces (Fig. 47a), but also on handsplit slate. PARKER (1942) first attracted attention to this prominent feature along a stone wall in Owego, N.Y. Later, ROBERTS (1961) observed feathers in homogeneous fine-grained rocks on open joint surfaces. HODGSON (1961) reports that breaking of such joints marked with plumes initiates

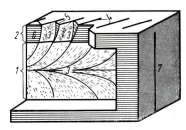

Fig. 46. Principle of feather markings on systematic joint surfaces

    1. Main joint face
    2. Fringe zone, only prominent on large surfaces
    3. Plumose structure as barbs or feathers
    4. Border planes (joints on fringe)
    5. Cross fractures, continuation of plume into fringe
    6. Shoulder of fringe to main joint face
    7. Trace of main joint face

From HODGSON (1961), modified from ROBERTS (1961)

Fig. 47a. Plumose markings on dolomite. Stone was probably broken for the purpose of dimension stone resulting in tension joints. The photo was kindly supplied by Professor J. M. PARKER III, and was also used by PARKER (1942), MUEHLBERGER (1961), and BADGLEY (1965)

at a point and propagates outward through the rock from the point of start. Plumes appear to start at the chisel point and spread outward like spiderwebs. Fig. 47b pictures plumose markings induced by chisel separation in Virginia slate. PARKER (1969) has shown evidence that plumose markings result from tension or extension as the antithesis of shear.

Presentation of Joint Directions

The directional distribution of joints as joint rosettes is often seen in a tectonic analysis of areas, but also in the analysis of a very small area. STINI (1925) plotted joint rosettes for both the compass directions, or azimuths, and the joint dip. STINI's original concept of the joint rosette is reproduced in a modern version in

Fig. 47 b. Feathering (plumose structure) on slate surface. Structure starts from chisel impact points (arrows). Virginia Buckingham slate

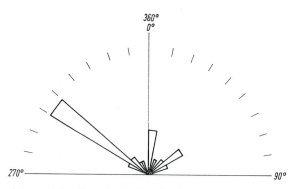

Fig. 48. Rosette of joint strike lines in the Poschach granite quarry, Mauthausen, Austria. The prominent northwest-southeast strike direction reflects the main course of the Danube river along the faulted zone of the south edge of the Bohemian Massif. Rosette drawn from data of STINI (1925)

Fig. 48, for a limestone quarry in Austria. The high accuracy of the joint strike presentation permits its application to quarry operations. For very critical work, the joint characteristics can contain information about joint fillings, repaired joints, etc.

Faulting

A fault is a fracture or a fracture zone along which displacement of one side has occurred relative to the other. The displacement may be a few inches or many miles. Faults may have a variety of effects on stone and the stone production, depend-

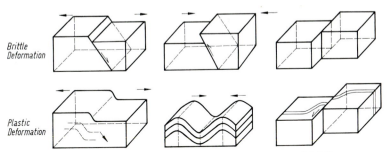

Fig. 49. Principles of rock deformation, faulting and folding
a. brittle deformation: tension leads to normal faulting; compression leads to reverse fault-
    ing; shear leads to strike-slip faulting
b. plastic deformation: tension leads to flexures, monoclines; compression leads to folding
    both symmetric and asymmetric; shear leads to drag folding

Fig. 50. Faulting on boulder of banded gneiss in glacial drift. Two different generations of
repaired faults are visible:
1: older fault by tension (normal fault), with about one foot displacement
2: younger fault by compression (reverse fault), with about three inches displacement

ing on the size of fault, the kind of fault filling and the degree of rock shattering in the immediate vicinity of the fault. Soft fault gouge and weathering along faults has often complicated quarry operations and rendered stone worthless.

Fig. 51. Crushed fault zone with minor movement. Arrows mark the possible direction of movement. Fault also crosses shear zone with fractures concentrated near shear zone. Cracks and fault are filled with graphite in white, fine-grained marble from Alaska. Store-front veneer, Colorado Springs, Colorado

Fig. 52. Complexly faulted drag-folded marble in column, bandeed green-and-white Cipollino marble from Egypt. The generations of movement are visible:

1. Major normal faulting with downward drag of marble
2. Subsequent reserve faulting (compressional with backdrag)
3. Minor downward drag along fault, some breakage of band

Column to the right edge is rectangular and veneered in diamond pattern. Vatican Museum

Crushing along a fault may merely be replaced by polished and striated surfaces which result from movement along a fault plane, called slickensides. Striated calcite filling along a fault is a common feature on a fault plane which was filled with white calcite, quartz or some other mineral substance. In quarrying, slickensides provide a natural line of separation and a specially patterned surface for dimension and flagstones.

Cemented (repaired) fault zones occur where crushing was minor and complete cementation has occurred. Boulders of such faulted granites and gneisses are desirable ornamental objects to landscapers.

Clay fillings along fault zones form a distinct line of separation which often mean a great danger or at least a nuisance in quarry operations. Weathering processes readily enter along faults and deteriorate the rock rapidly towards both sides of the fault. Faulting through compression and through tension is illustrated in Fig. 49, with its relationship to folding. Examples of faulting and folding in stone are pictured in the Figs. 50 and Fig. 52 combines faulting in marble with pronounced dragfolding.

3.2. Plastic Deformation

Plastic deformation of stone slabs often annoys builders because warping and buckling on buildings may occur. A thin slab of crystalline marble supported on both ends of the slab gradually tends to bend through the unsupported midsection

Fig. 53. Plastic deformation of marble slab under own weight. The high relative atmospheric humidity has accelerated the process. Cemetery, New Orleans, Louisiana

under the own load of the slab. This plastic flow takes place as intergranular movement on glide planes or as gliding along twinning planes. Moisture acts as a catalyst rather than a lubricant. Many marble slabs on the graveyards of humid New Orleans, Louisiana, have been strongly deformed and distorted by plastic flow (Fig. 53). Plastic deformation may overlap with deformation by stress relief.

Folding

Plastic deformation may lead to folding (see Fig. 49) when compressional, tensional or shear processes proceed slowly enough for plastic flow to develop. Folding may be observed in all rock types, especially on metamorphic marbles and gneisses; in these rocks, plastic conditions generally prevail at their depth zone of formation. Combinations of folding and faulting are common, marking several generations of rock deformation on a single stone block or column (Fig. 52). Flow-folding can grade from true metamorphic rocks into igneous rocks where in some granites "primary flow-folding" from a slowly moving crystal mash has created strange flow structures with occasional inclusions from the wallrock. "Rainbow"-type granitic rock probably flow-folded in the liquid-plastic state in contrast to true metamorphic flow-folding of marbles in a solid plastic state. Many different types of fine-grained Vermont marble and other marbles feature very interesting combinations of folds, flow-folds, dragfolds, etc. on decorative veneer slabs and graveyard headstones. A further discussion of the complex mechanism of faulting and folding is beyond the scope of this book. Small stone slabs or stone blocks do not readily disclose complex mountain-building processes, as they lack a detailed perspective in space and time.

3.3. Active Rock Pressure in Quarries

Minor rock deformations in shallow quarry operations often take place as rock bursts or as relatively slow creep. Rock in its natural setting has established a temporary equilibrium with the external forces of erosion and weathering. Catastrophic degradation of very steep rock slopes towards more gentle slopes represent progress toward a natural equilibrium; an open pit quarry defies man's interference with often violent reactions: it tends to re-establish equilibrium by the redistribution and reorganisation of stresses. The response may take hundreds of years — or only minutes. A vertical cut by saw or torch can close while cutting is still in progress.

Rock pressures in quarries are generally ascribed to mainly two sources, stress relief by eroded overburden and true mountain-building tectonic pressure.

Stress Relief

Stress in rock is in equilibrium in the interior of a solid when neither normal nor shear stresses are transmitted through its surface (VOIGHT, 1966). Such stresses can be of two types:

a) Stresses which arise in a material through inhomogeneity of external stresses. Residual stresses develop within a plate if the plate is bent and some plastic deformation prevents regaining of the original shape.

b) Internal stresses can develop by minor defects of the grain properties through prestressing during previous mountain-building events (VOIGHT, 1966). The removal of confining stresses during the quarry operation gradually leads to

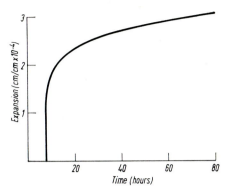

Fig. 54. Volume expansion of marble with time. After Voight (1966)

Fig. 55. Sheeting in granite, well developed parallel to the terrain surface, diminishing downward. Hayward Granite Quarry, Graniteville, Missouri

Fig. 56. Sheeting in granite forms natural horizontal lines of separation (arrows). Note different surface texture by flame-cutting, wire-sawing, and wedge-splitting. Marble Falls granite quarry, Marble Falls, Texas

expansion towards its original prestressed condition. The expansion of prestressed marble, granite or gneiss may take years to recover moving fast at first and diminishing in velocity with the passage of time. VOIGHT's data are plotted in Fig. 54 for 80 hours of expansion. KIESLINGER (1960) analyzes residual stress in a granite quarry on both a bare rock surface and weathered rock mantle, as well as the influence of stress relief on sheeting which reaches from the upper edge of the quarry downward to about 150 feet. Stress relief developed along distinct horizontal joints which lie approximately parallel to the original land surface when unloading of the overburden by erosion began some time in the geologic past. Sheeting intensifies towards the land surface to form major exfoliation surfaces and exfoliation domes (Figs. 55, 56). Natural stone cleavage in igneous rocks is the result of an invisible preferred orientation and direction of weakness; stone

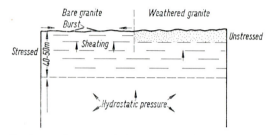

Fig. 57. Stress distribution and stress relief near the terrain surface, in unweathered and weathered granite. After KIESLINGER (1960)

Fig. 58. Large rock burst (bump) and separation of thin sheets of granite. Mt. Airy granite quarry, North Carolina

cleavage eases separation of large blocks by wedging. Occasional horizontal stress relief may cause considerable compressional forces, leading to bumps and rock bursts. Numerous rock bumps and the separation of thin sheets are common in the Mount Airy Granite quarry near Winston-Salem, North Carolina (Figs. 57, 58).

The slabs are only a few inches thick and are often of considerable length; the height of the rooflike bumps is up to two feet in the center. Sheeting eases quarrying but also permits the entry of weathering agents. Stress relief on slabs of marble, occasionally on granite, is a common phenomenon. Buckling of marble slabs is observed more often than in other rocks as marble de-stresses exceptionally fast.

Rock may also be stressed and prestressed by mountain-building processes, i.e., tectonic pressures. Such stresses may be especially active near major fault zones. Quarry operations disturb the temporary stress equilibrium and re-activate the mountain's activity. Stress relief after tectonic pressure is generally directional. Extensive rock pressure developed along a major joint lineament and reacted violently in the Hayward Granite quarry of Graniteville, Missouri. The stone mined from this zone was weak and often crushed and the cutting of 2 to $2\frac{1}{2}$ inch trenches was unsuccessful because they closed behind the flame-cutter.

Stress-relieved stone blocks of granite in a quarry can increase in crompessive strength as they expand, often over a period of a few years (KIESLINGER, 1967, Fig. 59). Very strong stress relieving can lead to micro-cracking and unfavorable

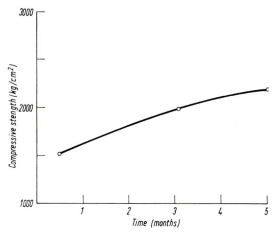

Fig. 59. Increase of compressive strength in granite with time, by recovery from pre-stressing. Data plotted from KIESLINGER (1967)

changes of some of the physical properties (OBERT, 1962). Strong winter freezing and the influence of heating accelerate the process of de-stressing. Stone blocks should be stored for a few months both for the purpose of curing (casehardening) and stress-relief; the process may be summarized as "aging". A full evaluation of stress conditions and directions of major stresses, both vertical and horizontal, are very difficult to estimate unless the evaluation of the quarry site is preceded by on-the-spot stress analysis; the degree of stress-relief behaviour can be very different for various regions. Details on methods, techniques and calculations of rock pressures are to be found in books on rock mechanics.

3.4. Quarry Operations and the Art of Stone Transport

The subject is generally well covered in handbooks on stone and on quarry operations. Some of these references are quoted in the chapter Technical Properties of Stone. Yet, a few words should be said on the art of quarry operation and stone transport.

Quarrying

Operational efficiency, slope safety, rapid and safe stone cutting without generation of vibrations, and the close observation of stone structures and rock stresses deserve much attention. Stone quarrying can range from cutting up large field stones, e.g., glacial erratics and residual granite boulders, to most elaborate underground mining techniques. Stone quarries and the art of quarrying is well discussed in the various mining journals, in Stone Magazine, Stone Industries, Pit and Quarry, and others. BARTON (1968) treats the practical aspects of the stone industry in detail, with many up-to-date references which will not be repeated here.

Stone Cutting in Quarries

Careful stone cutting — and some blasting — should separate the stone without micro-cracking. Advantage is taken of natural lines of stone separation such as bedding planes, joint surfaces and internal rock structures. Massive, unoriented granite usually has also a preferred direction of separation, or "cleavage" which the experienced stone worker finds readily (Fig. 5, chapter on Rock and Stone).

a) *Wedging:* The ancient wooden wedges are now replaced by steel wedges with "feathers", widely spaced and well aligned for splitting large blocks. Wedges are driven by hammer to a certain pitch till the block splits along predetermined lines through tensile forces.

b) *Drilling:* Closely spaced drill holes with often less than one inch of rock substance left between the drill holes is a popular method of cutting large blocks where helpful joints are not available. Rapid pneumatic or electric drills make this method economical today in locations where wire saws cannot be placed.

c) *Wire-saw cutting:* Endless braided steel wires are an inexpensive cutting tool in continuous operation both in soft and in hard rock.

d) *Flame-cutting:* or jet-cutting with a torch using a mixture of kerosene and oxygen was believed to be the inexpensive answer to all quarry cutting in the early 1960's. Not all rock, however, responds equally well to flame cutting. Today the method is primarily applied to igneous rocks where channels are cut by flaking and spalling, and by rock melting. Torch-cutting is the only possibility of stone cutting in areas where stressrelief occurs.

Stone transport of large blocks and monoliths over long distances has been achieved since the dawn of civilization. Large blocks can be readily handled today with modern cranes, giant flat cars and barges on waterways. Modern transportation engineers are puzzled by the past technology of moving stone blocks to 1000 tons in weight. Blocks for the circular Stonehenge of northwestern Europe were moved several hundred miles from their source where they were precision fitted to serve as astronomical computers. The largest block transported in Brittany weighs 382 tons. Incomplete projects or monuments left in the quarries have shed some light on the means of transport and subsequent erection. A very detailed account of transporting techniques is given by HEIZER (1966) who attempts to reconstruct the ancient techniques of giant-monument transport in an excellent well-illustrated summary documented by historical pictographs.

3.5. Damage to Stone by Blasting and Bombing

Much damage has been done to natural stone by unskillful blasting and by explosive bombs during World War II. The magnitude of the charge, the distance of the stone or structure from the explosion and the delay of the explosion after impact determine the degree and kind of damage.

Damage to Stone by Quarry Blasting and Explosive Bombs

1. Shock action: Stone fragmentation and micro-cracking by blasting and bombing can damage stone visibly or invisibly. Quarry blasting is intended to aid the economical separation and removal of large blocks where natural lines of separation are missing. For the recovery of broken stone, blasting is used to shatter stone sufficiently for removal. Most commercial dimension stone quarries avoid blasting entirely, to prevent unnecessary shattering and micro-fracturing. Both damage and blasting efficiency depend on existing joint surfaces, joint

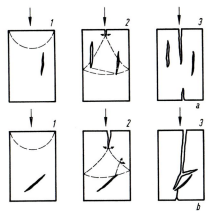

Fig. 60. Opening of fractures by an advancing blast wave. A Shock wave encounter is parallel to fracture. B Shock wave encounter ist at an angle of about 45°. From MUELLER (1964)

patterns, bedding or foliation, lithology and water content, according to MUELLER (1964). Detailed studies of rock blasting in quarries and tunnels were performed by DUVALL and ATCHISON (1957) of the U.S. Bureau of Mines. The actual shattering is believed to be based by the authors on the "Hopkinson Bar" principle whereby the reflective wave reflects as tension off a free surface, causing slabbing (at bedding planes, joint surfaces, etc.). The shock wave advances from the shot point with compressive stresses till the wave reflects back from any free surface towards the shot point as tensile stresses. Rock failure occurs, therefore, at relatively low stresses because the tensile strength of rock amounts to only a fraction of the compressive stresses. MUELLER (1964) illustrates in Fig. 60 the sequence of the mechanics of rock fissuring as extension and shear fractures by an incoming shock wave.

The shock effect on rock and stone can be expressed in terms of the strain energy, the particle velocity, and the shock pressure in kilobars. LANGER (1965) gives the approximate relationship between the particle velocity, the distance from the shot point and the size of the explosion in TNT equivalents. LANGER's three-dimensional graph appears to apply to shock safety for underground structures against explosion shock; the graph is modified for safety distances in stone quarries (Fig. 61). SHORT (1961) determined the maximum range of micro-fracturing in rock, 135 m. for a 500 kg. TNT explosive charge, more than 350 m.

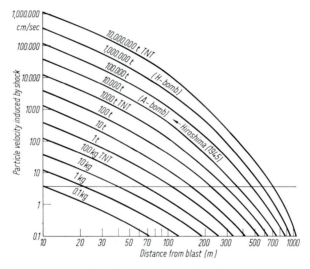

Fig. 61. Relations between particle velocity, magnitude of blast and distance from blast. Approximate line of quarry safety is marked with dashed line at 5 cm/sec particle velocity induced by shock. After LANGER (1965)

maximum distance for a 10 kiloton charge. A particle acceleration of 5 cm./sec appears to be still a safe shock velocity for most rocks used in the stone industry. Today most blasting is done with inexpensive ammonium nitrate. LEET (1954) shows that effective rock fragmentation and less vibration are generally achieved with short-period delay detonators in quarry blasting. SCHULZ and WUESTEN-HAGEN (1967) observed 25—30% less acceleration during careful, controlled blasting than with conventional methods; this results in less rock shattering. SCHWANENBERG (1967) summarizes the available information and techniques for blasting with minimum vibration. The following four modern smooth-rock separation techniques are summarized:

a) *Line drilling:* Serial perforation with 2—3″ diameter bore holes are spaced two to 4 times their diameter for maximum efficiency, without blasting.

b) *Cushion blasting* produces shearing along drill holes by buffered light, but complete charges in all drill holes.

c) *Pre-splitting* (or pre-shearing) is achieved by partial charges, often alternated with unloaded bore holes. The charges are set off before the main production blast.

d) *Smooth-wall blasting*, along one or two rows of bore holes, often fired after the main production blast, is related to cushion blasting.

Much research is in progress on the elimination of vibration during controlled blasting. Proper blasting is an art which requires extensive experience in the quarry. This art, however, can and will be quantified in the near future to save valuable rock from unnecessary shattering, and also to conserve the valuable real estate of rapidly expanding urban areas.

2. Splinter action: Splinter impacts on stone surfaces are shallow and conical, caused by shrapnel markings and by less-sharp stone fragments.

a) *Incendiary bombs:* High-temperature incendiary bombs were used extensively during World War II, resulting in major conflagrations from burning phosphorus. The behaviour of stone in fire is discussed in the chapter on Fire Resistance of Minerals and Rocks.

b) *Atomic bomb explosions:* Our knowledge of the effect of nuclear devices on stone dates from Hiroshima and Nagasaki (1945) and includes more recent model explosions in the USA. Heat and shock, both of very high intensity are generated for a very short time, but sufficient to alter the physical characteristics of rock. The damage to building materials in Hiroshima and Nagasaki was both by shock and heat. The effect was well analyzed by TESTER (1948), and in more detail by WATANABE *et al.* (1954) who describe and compare the blasts of Hiroshima with that of Nagasaki. One bomb exploded 570 m. above Hiroshima, the other 490 m. above Nagasaki, both were about 20 kilotons of TNT. The important natural building material was granite from at least 2 different localities. Mechanical shattering of granite blocks reached 8 to 10 inches beneath the stone surface. There is evidence that compressive crushing by vertical force exceeded 20 to 30,000 psi compressive strength. Roof tiles melted and stone fused at surfaces directed towards the blast in a radius of 1000 to 1600 m. away from the hypocenter, i.e., the projection of the bomb explosion to the earth's surface. Shock and thermal metamorphism were observed by JAMES (1969) after a buried atomic device was set off at the Nevada Test Site. The zone of intense change is characterized by basaltic glass and vesicular plagioclase glass; other minerals were intensely fractured. Further away from the blast the zone of moderate change still showed some plagioclase glass whereby other minerals were only micro-fractured.

An extrapolation and evaluation of the published data of the devastating surface effects of atomic explosions is very difficult to estimate. The type of damage to stone and other natural building materials by atomic and hydrogen bombs should be expected to be similar to the effects of smaller devices; the radius of stone damage, however, is much wider. The 5 cm./sec particle acceleration is probably valid for the estimate of the maximum range of rock micro-fracturing as well.

Summary

Jointing, faulting and folding are the result of former compression, tension or shear forces. Jointing and faulting form natural lines of stone separation in the quarry, but also may constitute weak and weathered zones or features of ornamental value.

Quarry operations are often impaired and delayed by rock stresses which do not become noticeable till the rock is tapped by man. Rock bursts and rock bumps, the closing of saw cuts should be ascribed to residual stresses caused by pre-stressing of rock masses. Pre-stressed stone slowly expands, de-stresses after mining: increase of volume, increase of porosity, increase of the compressive strength, but also the possible deterioration of the tensile and shear strength may be the result. Heating and freezing of the stone after mining accelerate the process of unstressing. Some cleavability of igneous rocks within the upper 150 feet below the present terrain surface is related to horizontal sheeting by near-surface stress-relief.

Damage to stone by blasting and bombing results in micro-fracturing. A tentative safety particle acceleration of 5 cm./sec is assumed by shock waves. A three-dimensional chart permits an estimate of the relation between blast intensity, particle velocity, and the distance from the blast.

References

1. BADGLEY, P. C., 1965: Structural and tectonic principles. New York: Harper and Row, 521 p.
2. BARTON, W. R., 1968: Dimension stone. U.S. Bureau of Mines, Information Circular 8391, 147 p.
3. DUVALL, W. I., and T. C. ATCHISON, 1957: Rock breakage by explosives. U.S. Bureau of Mines, Report of Investig., # 5356, 52 p.
4. HEIZER, R. F., 1966: Ancient heavy transport, methods and achievements. Science, **153** (**3738**), 821—830.
5. HODGSON, R. A., 1961: Classification of structures on joint surfaces. American Journal Science, **259**, 493—502.
6. HODGSON, R. A., 1961: Regional study of jointing in Comb-Ridge — Navajo Mountain Area, Arizona and Utah. Bull. Am. Assoc. Petroleum Geologists, **45** (**1**), 1—38.
7. JAMES, O. B., 1969: Shock and thermal metamorphism of basalt by nuclear explosion, Nevada Test Site. Science, **166**, 1615—1619.
8. KIESLINGER, A., 1957: Die Spaltbarkeit von Granit. Montan-Rundschau, **9**, 237—243.
9. KIESLINGER, A., 1960: Gesteinsspannungen und ihre technischen Auswirkungen. Z. Deutsch. Geol. Gesellschaft, **112** (**1**), 164—170.
10. KIESLINGER, A., 1967: Residual stress. Summary. Proc. of 1st Congress Intern. Rock Mechanics, **III**, 354—357.
11. LANGER, M., 1965: Die Dämpfung von Druckwellen im Gebirgskörper. Zeitschrift Zivilschutz, Baulicher Zivilschutz, **29** (**5**), 2—6.
12. LEET, D. L., 1951: Blasting vibrations effects. The Explosives Engineer, **28**,1 67—168; **29**, 12—16, 42—44.
13. LEET, D. L., 1954: Quarry blasting with short period delay detonators. The Explosives Engineer, 1954, Sept./Oct., pp. 142—154.
14. MUELLER, L., 1963: Der Felsbau, vol. I., Stuttgart: Ferdinand Enke, 624 p.

15. Mueller, L., 1964: Beeinflussung der Gebirgsfestigkeit durch Sprengarbeiten. Rock Mechanics and Engineering Geology, Supplementum I., pp. 162—177.

16. Obert, L., 1962: Effects of stress relief and other changes in stress on the physical properties of rock. U.S. Bureau of Mines, Report of Investigations 6053, 8 p.

17. Parker, J. M., 1942: Regional systematic jointing in slightly deformed sedimentary rocks. Geol. Soc. America Bull., **53**, 381—408.

18. Parker, J. M., 1968: Jointing in South-Central New York: Discussion. Geol. Soc. America Bull., **80** (**5**), 919—922.

19. Roberts, J. C., 1961: Feather-fracture and the mechanics of rock-jointing. Am. Journal Science, **259**, 481—492.

20. Schulz, H., und K. Wuestenhagen, 1967: Schonendes Profilsprengen beim Streckenvortrieb und im Tunnelbau. Nobelhefte, Sprengmittel in Forschung und Praxis, **33** (**1/2**), 12—32.

21. Schwanenberg, J., 1967: Ordnung, Merkmale und Anwendungsbereiche der Verfahren zum gebirgsschonenden Sprengen. Nobelhefte, Sprengmittel in Forschung und Praxis, **33** (**1/2**), 33—72.

22. Short, N. M., 1961: Fracturing of rock salt by a contained high explosive. Drilling and Blasting Symposium. Quarterly Colorado School of Mines, **56** (**1**), 221—257.

23. Short, N. M., 1966: Effect of shot pressures from a nuclear explosion on mechanical and optical properties of granodiorite. Journal Geophysical Research, **71** (**4**), 1195—1215.

24. Stiny, J., 1925: Gesteinsklüfte und alpine Aufnahmsgeologie. Jahrbuch der Geologischen Bundesanstalt, Wien, **75** (**1/2**), 97—127.

25. Tester, A. C., 1948: Effect of atomic bombing on building materials at Hiroshima, Japan. Geol. Soc. America Bull., **59** (**9**), 787—794.

26. Voight, B., 1966: Residual stresses in rocks. Proc. of 1st Congr. Intern. Soc. Rock Mechanics, pp. 45—50.

27. Watanabe, T., M. Yamasaki et al., 1954: Geological study of damages caused by bombs in Hiroshima and Nagasaki. Japanese Journal of Geology and Geography, Trans., **24**, 161—170.

4. Color and Color Stability of Structural and Monumental Stone

The color of structural and monumental stone challenges the engineer and architect to achieve with it the most effective and harmonious appearance in structural or architectural design. The utilization of different color shades of stone has given new life to many existing structures. Rock — or "stone" — is composed of one or more minerals. Stone colors are, therefore, influenced by the color of the predominant mineral itself, by the adjacent minerals, and by grain size.

In terms of their colors, the common rocks all have their own characteristics: the stable pigments of the igneous rocks (granites, etc.), the generally highly variable and frequently unstable pigments of sedimentary rocks (sandstone, limestone, and ornamental limestone-marble), and the quite different cold colors of the true metamorphic rocks (crystalline marbles, gneisses, slates, etc.). Each rock group is somehow delimited by a certain range of colors and textures, though the possibility of color variation appears unlimited. The architect should, therefore, be well aware of the different possibilities of color variations and pigment stabilities of commercial stones. To this end the attempt will be to summarize the available information on rock colors as well as on color stabilities.

4.1. Presentation of Color

A simple and unified presentation of color is of great importance to the properly color-conscious architect for purposes of accurate color description and comparison. Several systems of color comparison were developed by physicists; a brief summary is presented by WRIGHT (1969). Color may be regarded as psychological in category; but it is also possible to correlate physical properties with the psychologically determined attributes of color. In 1915 the original Munsell color System was published and has been frequently revised since. The result was a collection of color chips, the color of each being evaluated in this system in terms of hue, value, and chroma. The method of color notation developed by A. H. MUNSELL, as the principal feature of the system, arranges the three attributes of color into orderly scales of equal visual steps, so that the attributes become dimensions or parameters by which color may be analyzed and described accurately under standard conditions of illuminations for a normal observer viewing in daylight with grey to white surroundings; under these conditions hue, value and chroma of the color chips correlate closely with hue, lightness, and saturation of the color perception (the definitions for the color attributes are in part taken from the catalogue of the Munsell Company, Inc., Baltimore, Md.):

Hue: Chromatic colors in the Munsell System of Color Notations are divided into five principal classes which are given the hue names of red (R), yellow (Y), green (G), blue (B), and purple (P). A further division yields the five intermediate hue names of yellow-red (YR), green-yellow (GY), blue-green (BG), purple-blue (PB), and red-purple (RP), these being combinations of the five principal hues. The hues extend around a horizontal color sphere about a neutral or a chromatic vertical axis. When finer subdivisions are needed the ten hue names or symbols may again be combined to produce such combinations as red-yellow-red, which is symbolized as R-YR. For even finer divisions, the hues may be divided into ten steps each (1 R to 10 R). The designation 5 R marks the middle of the red hue, 1 R the faintest, almost grey, 10 R the strongest, deepest red. Further refinements

Table 13. *Hue Names and Abbreviations Used in the Munsell Color Chart*

Name	Abbreviation	Name	Abbreviation
red	R	purple	P
reddish orange	rO	reddish purple	rP
orange	O	purplish red	pR
orange yellow	OY	purplish pink	pPk
yellow	Y	pink	Pk
greenish yellow	gY	yellowish pink	yPk
yellow green	YG	brownish pink	brPk
yellowish green	yG	brownish orange	brO
green	G	reddish brown	rBr
bluish green	bG	brown	Br
greenish blue	gB	yellowish brown	yBr
blue	B	olive brown	OlBr
purplish blue	pB	olive	Ol
violet	V	olive green	OlG

From: NICKERSON Color Fan; maximum chroma, 40 hues. — Munsell Color Co., Inc., Baltimore 2, Maryland.

with "-ishes", e.g. reddish, complete the list of hue names as given in Table 13, which may be used with the unabridged Munsell Color Chart. Fig. 62 gives the location of the three attributes.

Chroma: The chroma notation of a color indicates the strength (saturation) or degree of departure of a particular hue from a neutral grey of the same value. The scales of chroma extend from /0 for a neutral grey, to /10, /12, /14, depending upon the strength or saturation of the individual color. Fig. 63 presents the purple section of the Munsell Color Solid to show the relationship of the color names.

Value: The value notation indicates the degree of lightness or darkness of a color in relation to a neutral grey scale, which extends in a vertical direction from a theoretically pure black, symbolized as 0/ at the bottom, to a pure white, symbolized as 10/ at the top. A grey, or a chromatic color that appears visually half-

Fig. 62. Hue, value and chroma in relation to color space. The values of white to black, and the chroma from grey to red are used as an example of the Munsell color presentation. From Munsell Book of Colors (1947)

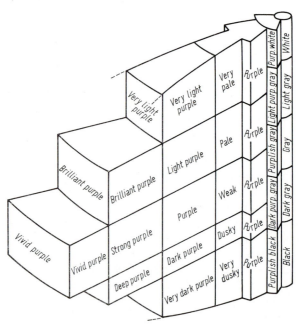

Fig. 63. Portion of color solid including purple, grey, black and white. From Rock Color Chart, GODDARD (1948)

way in lightness between pure black and pure white, has a value notation of 5/. Lighter colors are indicated by numbers above five, while darker ones are indicated with numbers below five. Table 14 correlates the ten Munsell Values with the commercial Kodak Grey Step Scale which is part of a Color Separation Guide for photographic purposes, and with the true light reflectance measured in per cent of daylight; this correlation should quantify the 10 Munsell value scale. The printed Kodak scales are commercially available.

Table 14. *Quantitative Comparison of the Munsell Lightness (Value) with the Kodak Grey Step Scale, and the Absolute Light Reflectance*

Value	Munsell System	Kodak Grey Step Scale		Light Reflectance
		scale values	calc. values	%
white	9	0.10	0.10	78.66
	8	0.20	0.23	59.10
	7	0.30	0.37	43.06
	6	0.50	0.52	30.05
	5	0.70	0.70	19.77
	4	1.00	0.92	12.00
	3	1.30	1.18	6.55
	2	1.60	1.51	3.13
black	1	1.90	1.92	1.21

From: Munsell System from A. H. MUNSELL (1947). Kodak Grey Step Scale from Kodak Notes on Practical Densitometry.

The complete Munsell notation for any chromatic color is written as Hue Value/Chroma. The color "medium red" should serve as an example for its presentation with its three color attributes: 5 R 5.5/6 : 5 R is located in the middle of the red hue; 5.5/ is the lightness of Munsell Value near the middle of light and dark; 6 is the degree of Munsell Chroma or color saturation, which is about in the middle of the saturation. Whenever a finer division is needed for any of the three attributes, decimals may be used such as 2.5 R 4.5/2.4.

The Geological Society of America selected the Munsell Book of Colors as the basis for its Rock Color Chart edited by GODDARD (1948) which is a valuable semi-quantitative attempt in determining the color with its three components by comparison. The Rock Color Chart is chiefly of value in describing the colors of medium to fine-grained rocks with 40 color chips. In describing very coarse grained rocks it is necessary to determine the color of each mineral, the chart is designed to cover their range. A blending of the individual colors can be secured by spinning the specimen like a color disc, or by looking at the rock from a short distance, and thus getting a monotone which can be matched with the hand-painted chips on the Rock Color Chart, which is commercially available.

The practical evaluation of the lightness of a colored rock surface would be done with the exclusion of the color perception; total color-blindness of the investigator would be helpful. The JUDD and WYSZESKY (1963) book on Color in Business, Science, and Industry provides more technical data on this subject.

Table 15. *Color Identification Chart for Common Rock-Forming Minerals*

Color	Mineral	Rocks		
		Igneous	Sedimentary	Metamorphic
red, pink	orthoclase	granite	sandstone (arkose)	gneiss
	hematite		sandstone	gneiss, marble
	calcite (rare)			marble
white	orthoclase	granite	sandstone (arkose)	gneiss
	plagioclase	granite	sandstone (arkose)	gneiss
	quartz	granite	sandstone	gneiss, quartzite
	muscovite	(granite)	sandstone, shale	marble, schist, quartzite
	calcite		limestone	marble
grey	plagioclase	diorite, gabbro		gneiss
	quartz	granite	sandstone	quartzite, gneiss
	graphite			marble, slate, gneiss
	organic substance		sandstone, shale, limestone	
black	hornblende	granite, gabbro, diorite		gneiss
	biotite	granite, gabbro, diorite		gneiss, marble, schist
	graphite			slate, marble, gneiss, schist
	organic substance		sandstone, limestone, shale	
brown	hornblende	granite, etc.		gneiss, schist
	biotite	granite, etc.		gneiss, schist, marble
green	hornblende	granite, etc.		gneiss, schist
	chlorite, sericite (muscovite)			gneiss, marble, schist
	glauconite		sandstone, shale, limestone	
	ferrous-ferric iron hydr.		shales, limestone, sandstone	slate, quartzite
yellow (ochre, buff, tan)	limonite minerals		shales, sandstone, limestone	
gold	sericite (muscovite)			marble
	limonite minerals		limestones	

The oversimplified Munsell Rock Color Chart has been found insufficient for more accurate color descriptions of rock and stone. FOLK (1969) therefore suggests the use of the available Soil Color Chart, published 1954, which is also based on the Munsell system, but which uses 248 color chips instead of only 40 with the Rock Color Chart.

The honing and polishing process of rocks decreases the light value through scattering and brings forth the hue by darkening the color. The value decreases also on a wet stone surface; therefore, a quarry-moist fresh granite surface may approximately suggest the proportion of hue, chroma, and value the stone will produce when polished. The value or lightness for each commercial stone should be presented to the customer. Darker colored stones reduce the value on a polished surface much more than light-colored varieties.

Rock colors are generally of a composite character caused by the blending of the different minerals, the size of the minerals and the particle cements. A great uniformity of the color is either due to the fine grain size of a single mineral or of evenly distributed pigment in a fine mineral matrix.

4.2. Colors of Igneous Rocks

Igneous rocks crystallized from a hot magma; these rocks may be either coarsely or finely crystalline depending on the rate of cooling. The following common rock-forming minerals help to determine the rock color: feldspars (orthoclase, plagioclase), quartz, micas (muscovite, biotite), ferromagnesium silicates (hornblende, augite), pyrite.

Orthoclase in granite and syenite: The common orthoclase feldspar may display colors from deep flesh to pink, to pure white; the great abundance of orthoclase in granites and syenites (50—75%) determines the color appearance of these rocks. Occasional crystals of white plagioclase may bring forth spottiness if the orthoclase matrix is other than white. The origin of the red hues of the orthoclase feldspars, although usually attributed to finely disseminated ferric oxide, is still unsolved. The plagioclase feldspars are not known to display red or flesh hues, presumably because iron cannot exist in them as ferric oxide.

Plagioclase of diorites and gabbros: The plagioclase feldspar group presents an isomorphous series from white, the sodium feldspar, through grey to the often black calcium feldspar. The plagioclase amounts to about 50% of all minerals in diorites and gabbros; it may also be a minor accessory in granites and syenites. The increasing iron content in darker igneous rocks apparently provides enough black ferrous-ferric iron pigment to result in dark grey to black plagioclases. The dark labradorite of some "black granites" may show iridescence which is ascribed to light refraction along the numerous narrow twinning planes. As individual crystals or vein fillings, all plagioclases are generally light in color.

Hornblende and Augite are common members of the large iron-magnesium silicate group with colors from green to black, rarely light green. The oxidation, and also further hydration of the ferrous-ferric iron during the weathering process, converts the minerals finally into a brownish or ochre-brown lustreless substance (see chapter on Natural Rust).

Biotite is dark brown to black in color as the result of the presence of ferrous-ferric iron, which is much more susceptible to weathering than iron in the other iron-magnesium silicates. Igneous rocks often contain biotite instead of hornblende or augite. The oxidation of iron in the silicate structure to ferric oxide or hydroxide strains the immediate surroundings of the mica to reddish-brown. The Biotite bleaches by loss of some iron to silvery white.

Muscovite is colorless to silvery grey and very stable in color. Granites and syenites may be rare hosts to this mineral.

Quartz of granites appears as small white or grey pearls, occasionally as blue opalescent grains which may produce a good contrast of colors. Quartz owes its white or grey color to entrapped air bubbles and minute imperfections of the crystal lattice, blue to the presence of titanium oxide.

Pyrite: All igneous rocks may contain pyrite in variable quantities. Relatively slow leaching and oxidation not only leaves ugly brown to ochre-brown stains but may also show halos of acid etching on adjacent mineral grains.

The generally unoriented mineral grains of nearly equal size give the stone a "salt and pepper" effect. This color fabric varies with grain size and mineral colors. Occasional dark-colored angular inclusions increase both the color contrast and the fabric. Flow-banded granites, e.g. the Minnesota Rainbow Granite, may be useful for many ornamental applications.

It may be summarized that the pigments of the different rockforming minerals of igneous rocks are quite stable. All minerals polish well and resist the deleterious influence of the city atmosphere, with the exception of biotite and occasionally pyrite.

4.3. Colors of Sedimentary Rocks

For the better understanding of the pigmentation of sedimentary rocks two major groups should be distinguished:

a) clastic sedimentary rocks: conglomerate, breccia, sandstone, shale;

b) chemical and organic sedimentary rocks: limestone, dolomite (non-cristalline marbles), onyx (calcite), alabaster (gypsum).

Many minerals of the sedimentary rocks are inherited from the parent igneous rocks, such as quartz, feldspars, hornblende, augite, mica, etc. Other rock-forming minerals, such as clay minerals in clays and shales, are formed by waethering processes; the color of such pure clay minerals is white or grey. The introduction of pigments such as iron and carbon to clays and shales, and also into sandstone and conglomerate, creates dark grey to black, red, ochre-brown, tan and other colors. The pigments of the sedimentary rocks can be summarized as follows:

Iron is the most common and the strongest pigment in sedimentary rocks. Most of the iron is precipitated having been either washed into the sedimentary basin as soluble iron or freed through the decay of organisms. Some of the iron may have been introduced during lithification, some during later secondary geological processes which are unrelated to lithification, such as mountain building.

Ferric iron may occur as the red hematite (Fe_2O_3), the reddish-brown to ochre-brown goethite (FeOOH), and the often amorphous $Fe_2O_3 \cdot n\, H_2O$; all these substances are common components of "iron rust". The color of most sedimentary rocks and soils is derived from various quantities of hematite, goethite, and amorphous iron hydroxide. A fraction of a per cent may be sufficient to add a warm brown, buff, tan, or red hue. Colloidal adsorption of iron hydroxide as thin coating on the surface of quartz grains is frequently found in many recent desert sand dunes as well as in ancient sandstones which are lithified sand dunes. The color of sandstone does not readily give information as to whether the iron was adsorbed on the grain surface or subsequently introduced with the cement. Breccias, conglomerates, and sandstones are often cemented with reddish-brown cement between the mineral grains; the contrast greatly enhances the ornamental effect of the stone.

Red limestone is rare; its beauty and inhomogeneity improves the rock to a highly priced ornamental stone. GALLOWAY (1927) considers the color of all red limestones as secondary, with the pigment introduced after the formation of the limestone. DIMLEY (1963) has found evidence that red limestones may be also formed in shallow lagoons with sufficient influx of red, lateritic soil from the former shore line as seems to be the case with some reddish Tennessee limestone-marbles; if the pigment had been introduced during catastrophic events, the scars of former crushing should be still visible. Lithification processes often attacked originally angular limestone fragments (breccias) by solution and converted these to rounded, poorly defined spherules with a color saturation less intense than the color of the cement subsequently introduced; the Italian Verona Red limestone-marble should be cited here as an example of a common veneer stone. Strongly pigmented limestones become more beautiful if cut by white veins. Veins of ochre-brown or red in a grey or buff limestone may occur where extensive solution had attacked the rock some time in the geologic past, filling the channels and small cavities with red residual soil which subsequently lithified. Some dark-grey limestone-marbles from West Germany with red cavity fillings trim the lobby of the Empire State Building in New York City. Although lithified, the clay fillings still may show up as weak spots which often have to undergo special treatment.

Ferrous iron may occur as finely distributed pigment, as the yellow-brass pyrite or marcasite (FeS_2) or as the grey or white iron carbonate siderite ($FeCO_3$).

a) Ferrous sulfide, amorphous or crystallized, appears to be black but shows a yellowish brass color if recrystallized to larger grains of pyrite or marcasite. Limestones and shales often contain minor quantities of deep-black, finely disseminated ferrous sulfide which was formed in a reducing environment. TOMLINSON (1916) demonstrates the relationship of ferrous to ferric iron in slates and its influence on the stone color (Fig. 64). Shales have similar color proportions as slates. Grey or green reduction centers are occasionally found in both red shales and red limestones. KELLER (1929) found by both observation and experiment that ferric oxide is reduced by H_2S to ferrous iron, after which it is easily soluble in waters charged with CO_2. The colors of shales, slates, sandstone, and limestone may be influenced by the shift of the proportion of ferrous to ferric iron. KIESLINGER (1964) demonstrated with chemical analyses on Austrian red limestone-marble from Adnet that the red and grey components contain the same amount of total iron, whereas the ferrous to ferric iron proportion changes only slightly, as illustrated in Table 16.

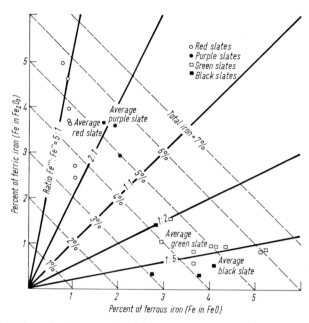

Fig. 64. Effect of the ratio of ferrous to ferric iron on the color of slates. The amount of total iron is the same in red as in black slates. After Tomlinson (1916)

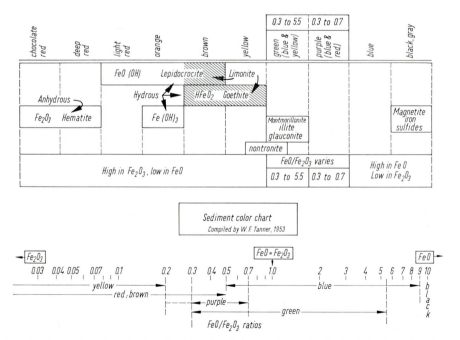

Fig. 65. Tanner's sediment color chart. Rock colors as related to the FeO/Fe_2O_3 ratio. From Hughes (1958)

W. F. TANNER has compiled a good summary chart of sedimentary colors brought forth by different pigments of iron (Fig. 65). GARREL's (1965) Eh-pH diagram of the ferrous and ferric iron stabilities in some important natural

Table 16. *Proportion of Ferric to Ferrous Iron in Red Adnet Limestone-Marble from Salzburg, Austria; the Red Color is Reduced to Grey Along Veins and Cracks*

Color	Total Iron %	Iron-III	Iron-II	Fe-III:Fe-II	Fe-II:Fe-III
Red part (10 R ¾-10 R 4/6)	0.22	0.09	0.13	1:1.5 = 0.6	1.44
Grey part (5 G 6/1-5 GY 4/1)	0.22	0.06	0.16	1:2.5 = 0.4	2.66

Data from: KIESLINGER (1964), p. 159. Color description as Munsell color values from Rock Color Chart.

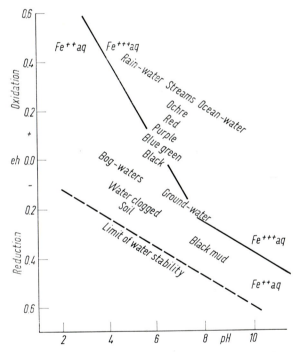

Fig. 66. Colors of sedimentary rocks as they are influenced by the pH and the Eh (Redox Potential) of the environment. After GARRELS and CHRIST (1965), and HUGHES (1958)

environments is reproduced in Fig. 66. The boundary line between the ferrous and ferric iron in aqueous solution is well reflected in the distribution of the colors for depositional and postdepositional environments of sedimentary rocks. With the help of Fig. 66, the conditions of Table 16 may be readily applied to a Verona-Red limestone-marble with green bands a long cracks in the following way: The marine limestone was probably originally deposited under reducing conditions, was later oxidized during crushing by mountain building processes to bright red, followed by submergence to below groundwater level where reducing waters penetrated along cracks, reducing the red ferric iron to green ferrous compounds adjacent to the channels. In contrast to the Verona-Red, the depositional sequence is reversed in the light pale-green Mexican Pedrara onyx-marble which is cut by many gold-brown veins and veinlets. Ferric iron solutions entered along bedding planes and cracks, creating a beautiful contrast with the original pale-green calcitic matrix; transmitted light displays a spectacular array of color patterns and shades for ornamental purposes.

b) Ferrous carbonate is an occasional component in limestone and crystalline marbles in very small quantities. Generally colorless or light buff, the iron carbonate dissolves in pore water easily and may be thus transported outward to the stone surface where it oxidizes to form yellow tarnish of limonite in blotches after the pore water evaporated off the stone surface. KIESLINGER (1949) cites the Pentelian marble of the Acropolis of ancient Greece as a good example for this type of surface yellowing, possibly through the oxidation of finely distributed iron carbonate. The staining is particularly well developed along outer edges, pinnacles, etc. The outward migration of capillary water prefers travel toward edges; the process may be compared with the migration of lime to the outermost edges of vaporizing plates of a furnace humidifier or a porous clay flower pot (see chapter Moisture and Salt Action).

c) Glauconite, a hydrous K, Fe, Mg, Al, silicate. Different shades of green may be occasionally found in limestone as bands or evenly distributed grains in sandstone and shale. Green sandstone is not rare. The mineral weathers to ferric compounds.

Iron pigments in sedimentary rocks are generally unstable if exposed to light and weathering; the degree of the color stability is generally hard to predict. As an example, the different colored varieties of Minnesota Mankato stone range from bluish grey to gold-buff. Field observation on the bluish-grey Mankato Grey Ledge could not detect notable color changes towards brown or ochre while the stone was exposed to oxidizing environments. The stone producer, however, admits that slight color change of the tan-colored Mankato Buff Ledge should be expected towards a golden-buff tone. This slow color change is being observed at the exterior of the new Notre Dame Memorial Library in six years of exposure to highly oxidizing conditions. Strongly pigmented stone is therefore not recommended for exterior use, especially if water soluble, because the direct sun light and the possible solubility of the stone quickly bleach the beautiful colors and convert them to dirty greys within a few years of exposure.

Carbon: Organic substance is frequently present, disseminated in fine-grained sediments. The color induced by the organic substance ranges from grey to deep black. FORSMAN and HUNT (1958/59) distinguish three types of organic matter in rocks on the basis of their chemical composition:

a) Hydrocarbons, pure solvent-soluble organic matter composed of carbon and hydrogen only.

b) Asphalt belongs to the group of solid and semisolid hydrocarbons largely soluble in carbon disulfide.

c) Kerogen, insoluble in solvents, which constitutes the bulk of the organic substance in shales and carbonate rocks; kerogen yields oil when the shales undergo destructive distillation. The carbon pigment of marine shales and limestones is closely related to the composition of coal: its hydrogen-to-oxygen ratio resembles that of marine shales and limestones. HUNT (1961/62) calculates the average proportion of the three different types of organic materials enclosed in shales and limestones, presented in Table 17.

Table 17. *Color and Average Organic Content of Finegrained Sedimentary Rocks*

Rock	Hydrocarbons (ppm)	Asphalt (ppm)	Kerogen (ppm)	Color
Shales	300	600	20.100	grey to black
Carbonates	340	400	2.160	grey to black
Shale (red and green)	15	40	1.000	red and green

From: HUNT (1961/62).

HUNT's classification of organic matter corresponds with BAKER and CLAYPOOL's (1970) observation, that the generally aromatic hydrocarbons determine the dark colors of sediments. The pigment stability of each carbon compound determines the color change to be expected during weathering or mild metamorphic processes, the hydrocarbons being the least stable carbon pigment.

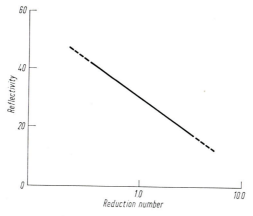

Fig. 67. Relationship of light reflectivity to reduction number, a measure of the content of organic substance of shales and limestones. Adapted from PATNODE (1949)

PATNODE (1941) presents evidence that a relationship exists between the organic matter content, and the degree of light reflection (Fig. 67).

Some grey carbonate rocks, such as limestone and dolomite, may release hydrogen sulfide gas spontaneously if hit with a hammer or if blasted. The color of some rocks may change to pure white as the gas escapes from along the mineral grain boundaries.

Dark-grey or black limestone-marbles are well liked for interior decoration and for sculptures since Baroque times. White fossil cross sections and also white calcite fracture fillings make the stone still more desirable as an ornamental stone. Dark carbonaceous limestones are found often in the vicinity of coal beds; the famous Bleue Belge marble of Belgium is a well known example. A high carbon content of sandstones may reduce strength to below the permissible minimum. Dark-grey and black limestone-marbles bleach quite easily to a dull, dirty light grey within a few years of exposure to weathering agents. Solution also attacks the polished stone surface and increases the lightness or value from a dark towards a very light tone.

4.4. Colors of Metamorphic Rocks

Gneiss, schist, slate, quartzite, and crystalline marble are the most frequently used metamorphic rocks.

Gneisses and most *schists* are composed of minerals whose stable pigments are mostly inherited unchanged from their parent igneous and sedimentary rocks; the colors are therefore similar to those of igneous rocks. A few new important pigment minerals are added to the inherited minerals, such as sericite (fine-grained muscovite), chlorite and graphite.

Sericite is a very fine-grained silvery to greenish muscovite which is a frequent accessory mineral of many marbles and schists.

Chlorite owes its green color to hydrated ferro-ferric iron oxide.

Graphite is a flaky grey to silvery-black mineral found in many marbles, slates, and schists, and is metamorphosed carbonaceous substance. All minerals except graphite are quite color-stable pigments.

The schistose or foliate structure gives these rocks a pronounced oriented pattern. Strong foliation of most metamorphic rocks, however, rarely permits their use as structural stone.

Sedimentary limestones, conglomerates and sandstones undergo a considerable color change during metamorphism, in contrast to the pigment-stable igneous rocks. Pigment minerals recrystallize much as do other minerals, and concentrate to form larger grains conforming with the grain size of the rock matrix. Frequently, the pigment mineral may occur as occasional strongly concentrated bands, sometimes forming banded marbles of solid green or black.

Formerly strongly carbonaceous limestones are metamorphosed to marbles in which carbon is concentrated as graphite in bands and along joints.

Warm tones of tan, buff, cream, ochre, or black in the original sedimentary rock change to a cold white with green, silver, yellow, or dark grey color. A finely distributed limonite of a tan limestone-marble will yield a cold-white marble with a few scattered tiny black grains of hematite or magnetite; a dark-grey or black limestone with carbonaceous pigment may change to a white marble with black graphite flakes arranged in clusters or bands, or to an all-white marble if the pigment burned out. BAIN (1934) showed the transition of bituminous, black fossiliferous limestone of northern Vermont, the commercial Radio-Black, into white crystalline marble near Proctor, Vermont, the Vermont Marble of the Vermont marble-slate belt. Hydrocarbons and pure asphalt frequently burn out during the process of metamorphism; this obviously has happened in the case of the Vermont Marble. The proportion of kerogen to soluble asphalt and pure hydrocarbons plays an important role in the understanding of the pigments in metamorphic rocks.

Light pink and red crystalline marbles are rare and the source of their pigment is often not clear, as is the case with the Georgia pink marble (Etowah Pink) and Norwegian Rose. Red and pink quartzites, in contrast to pink marbles, are quite common; red hematite is the powerful pigment in the generally finer grained quartzites. The different varieties of the Tennessee marbles are not true crystalline marbles, and are therefore discussed with sedimentary limestones. Table 18 summarizes the pigments of both the sedimentary and metamorphic rocks.

Table 18. *Metamorphism of Pigments in Sedimentary to Metamorphic Rocks*

Sedimentary Rocks		Metamorphic Rocks	
pigment	color	pigment	color
limonite (goethite)	ochre, brown, tan, cream	hematite	red, bluish black
hematite	red to reddish brown	magnetite	black
soluble hydrocarbons	grey to black	burn out	
insoluble kerogen	grey to black	graphite	black to silvery grey
clay, as impurity in limestone and sandstone	shades of grey and green	muscovite (sericite) chlorite	silver, white, gold light or dark green

Pigment stability in metamorphic rocks: Most pigments of the metamorphic rocks are considered quite stable if the rocks are exposed to light and weathering. The pigment stability of polished crystalline marbles should not be confused with the increase of the lightness of polished rock surfaces caused by the rapid loss of the surface polish through solution by atmospheric agents during exterior exposure.

4.5. Color Variations in Stone

Weathered joint surfaces (seamface): Different color hues and chroma may be obtained from a single type of stone from the same quarry if surfaces with different ages of weathering can be produced. Facing stone from along open joints of granites, for instance, can present color varieties from light tan to ochre and dark brown, where weathering has progressed differently along narrow natural channel ways. Thus, facings of freshly broken stone (splitface) properly alternated with weathered faces may offer a lively colored surface which is pleasant to look at.

Discolored joint surfaces are only superficially weathered; the general strength of the stone is not reduced (Fig. 68, 69 Farbtafel).

Color combinations of the same stone are generally subtle and desireable to the architect, whereas the use of entirely different stone intermixed along a single surface has to be done with extreme care as a wild assortment of colors may become quite offensive.

Liesegang rings: In 1896 Liesegang explained rhythmical concentric banding in agates through experiments with the diffusion of silver nitrate in gelatin. GORE (1938) demonstrates that iron hydroxide may form Liesegang Rings by diffusion through a non-gelatinous medium as well. Orange and ochre-brown concentric

Fig. 68. Six different shades of granite from Quincy, Mass., from grey to brown set into a single wall: seam and splitface; entrance to the Battery Tunnel, Manhattan, New York

Description of colors using the notation of the Rock Color Chart:

N 7	light grey	10 YR 6/4	moderate yellowish-brown
N 5.5	medium grey	10 YR 3/3	dark yellowish-brown
5 YR 5/5	moderate yellowish-brown	10 YR 5/3	moderate yellowish-brown

rings of different thickness are often found in sandstone, and also in weathered igneous rocks. Larger field stones often show such concentric rings which are roughly parallel to the original stone surface or parallel to the original location of open joints Fig. 69a. More details about Liesegang Rings are given in the chapter on Moisture and Salt Action.

Stylolites (crowfeet) whith dark irregular zig-zag lines run approximately parallel to bedding planes of some soluble sedimentary rocks; they may bring forth strange color effects on surfaces of decorative stone. The green or black clay fillings along the stylolites, often thicker near and at the apices of these lines, form distinct color and fabric patterns against the differently colored limestone. A large number of Tennessee marbles should be cited here as good examples for stylolites as desirable architectural features (Fig. 16).

Dendrites: Dendrites are branch-like thin coatings of bluish-black iron-manganese oxides which crystallized from solutions along joints and bedding planes resembling fine, moss-like formations or fern-like organic imprints. Dendrites along bedding planes of Mexican onyx (Aztec onyx) form bluish-black moss-like coatings on bedding planes and contrast with the honey-yellow translucent onyx-marble

(Fig. 69b). Dendrites can enhance the beauty of any stone surface, but can also mar a surface with crude designs.

Special color effects: Sedimentary carbonate rocks frequently show local beds with accumulations of darker or lighter colored shell cross sections. Coral branches

a   b

Fig. 69a. Yellowing of granite along joints and near surface of former field stone

Fig. 69b. Black Virginia slate with uneven cleavage, set in "crazy-pattern" is contrasted against red New York slate with even cleavage on the walkway. White ¼″ sized quartz gravel fills the joints between the red slate. Mexican "Golden Cave Onxy" panels transmit daylight to both sides of entry to residence

Fig. 70. Liesegang rings of limonite across bedding planes of Briar Hill sandstone of Ohio

with different orientation and a gradually changing matrix may create a variety of irregular patterns and colors in large surfaces or ornamental panels; the corals may be white in a darker grey, red or black matrix, or the matrix may be light

6a*

grey and the branches colored. The stone manufacturer should point out unexpected color irregularities to the customer as sample slabs may become meaningless. In this case the architect may have to select the panels himself if he seeks to obtain certain combinations of texture and color. Metamorphic marbles are more consistent in color and fabric than sedimentary marbles.

Stone colors for mosaics: Mosaics of differently colored stone have been assembled since ancient time. The Notre Dame Memorial Library stone mosaic (see color plate Fig. 71) adorns the south face of the library, 135 feet high and 65 feet wide. Eighty-one different kinds of stone with different colors and finishes in a total of 5579 chips offered both the desired hue-chroma-lightness proportions and also special effects through the application of different rock textures (WINKLER, 1966). The pigment stability of each component had to be well known and assurance given that later color changes would not disturb the well-planned balance laid out by the artist.

Summary

Color is defined with three components, hue, chroma, and value or lightness; the components may be plotted in three dimensions. The overall rock color depends on the mineral colors and on the rock texture and structure. Fine-grained rocks appear to be more homogenous than coarser grained varieties. The process of polishing and honing darkens the stone, it decreases its value without changing its hue or chroma.

Igneous rocks, granites, syenites, gabbros, basalts and others, bring forth colors of red, pink, white, and grey with transitions to black. Occasional inclusions and flow banding offer a variety of patterns. All the pigments of igneous rocks are

---

Fig. 71. Notre Dame Memorial Library; with Stone Mural Mosaic
Natural stone is the major building material for facing which had to be delicately balanced against the unique stone mural mosaic, against the brick of the lower two stories, against the exposed aggregate of the concrete walkways and against its architectural environments. Mankato Golden Ledge dolomite features the entire exterior of the library tower and rims the brick wall of the lower two floors.
The stone mosaic was assembled after Millard Sheets' Word of Life, with Christ as the teacher surrounded by his apostles and assembly of Saints and Scholars who have contributed to knowledge through the ages.
The mural is composed of 324 panels of which 189 are pre-cast panel units. The balance of 135 are solid granite and Mankato Golden Ledge Stone as background panels. 81 different materials in a total of 171 finishes were used in fabrication. The cast panels consist of 5,579 individual pieces. The inclusion of the solid panels brings the count up to a grand total of 5,714 pieces. The largest panel weighs 3,218 pounds with a size of 5′ 7/8″ by 10′ 11/16″. The head of Christ, cast as one unit, contains the highest number of pieces, 115. The 81 different stones of different surface finish from 16 countries around the globe are of the following geologic origin: 46 granites and syenites, 10 gabbros and labradorites, 4 gneisses, 12 serpentines, 4 marbles and 5 limestones. The stone colors, fabrics, structures, and surface finishes give the desired effect.
The color photo was supplied and the color plate donated by Ellerbe Architects of Saint Paul, Minnesota, the architect of the Notre Dame Memorial Library

Applied Mineralogy 4, Winkler, Stone

Fig. 71. Notre Dame Memorial Library (see p. 84)

very colorfast, enough to last a few human generations under even adverse climatic conditions.

Sedimentary rocks, conglomerates, sandstones, limestones, dolomites, and shales are rocks which contain pigments of red ferric oxide, buff ochre, tan ferric hydroxide, or green and grey ferrous iron. Organic compounds produce color ranges from grey to black. All pigments of sedimentary rocks, especially of the limestones, are characterized by warm color tones, in contrast to true crystalline metamorphic marbles which display cold tones. All their pigments are finely distributed throughout the rock, and are usually not colorfast. Water-soluble carbonate rocks fade faster through their increase of the light value.

Metamorphic rocks, gneiss, crystalline marble, slate, and quartzite, have generally stable pigments, which may be arranged as individual grains, or as colored bands of yellow, gold, green, silvery grey, and black. Newly formed micas color rocks silver, gold, brown, and various shades of green; graphite contributes grey to black. The pigment of most metamorphic rocks are more stable than the pigments of sedimentary rocks.

Special color effects may be obtained through:

1. variation of tone in facing stone through the use of tarnished surfaces;

2. concentric diffusion rings (Liesegang Rings) cross-cutting bedding planes of sandstones;

3. "Crow-Feet" (stylolites), well marked zig-zag lines in limestones;

4. cross sections of fossils, cavity fillings with red residual soils, and others.

References

1. BAIN, G. W., 1934: Calcite marble. Economic Geology, **29** (**2**), 121—139.
2. BAKER, D. R., and G. E. CLAYPOOL, 1970: Effect of incipient metamorphism on organic matter in mudrock. The Am. Assoc. Petrol. Geol. Bull., **54** (**3**), 456—468.
3. DIMLEY, D. L., 1963: The "red stratum" of the Silurian Arisaig series, Nova Scotia, Canada. J. Geology, **71** (**4**), 523—524.
4. FOLK, R. L., 1969: Toward greater precision in rock-color terminology. Geol. Soc. America, Bull., **80** (**4**), 725—728.
5. FORSMAN, J. P., and J. M. HUNT, 1958/59: Insoluble organic matter (kerogen) in sedimentary rocks. Geochim. et Cosmochim. Acta, **15**, 170—182.
6. GALLOWAY, J. J., 1927: Red limestones and their geologic significance (abstract). Bull. Geol. Soc. America, **33** (**1**), 105—107.
7. GARRELS, R. M., and C. L. CHRIST, 1965: Solutions, Minerals, and Equilibria. New York: Harper and Row, 450 p.
8. GODDARD, E. N., *et al.*, 1948: Rock Color Chart. Geol. Soc. America, Natl. Research Council; republished by Geol. Soc. Am. 1951.
9. GORE, V., 1938: Liesegang rings in non-gelatinous media. Kolloidzeitschrift, **32**, 203—207.
10. HUGHES, R. J., 1958: Kemper County geology. Mississippi State Geol. Survey, Bull. # 84, 274 p.
11. HUNT, J. M., 1961/62: Distribution of hydrocarbons in sedimentary rocks. Geochim. et Cosmochimica Acta, **22**, 37—49.
12. JUDD, D. B., and G. WYSZESKI, 1963: Color in business, science and industry. New York: John Wiley, 500 p.
13. KELLER, W. D., 1929: Experimental work on red bed bleaching. Am. J. Science, 5th ser., **18**, 65—70.

14. KELLER, W. D., 1962: Clay minerals in the Morrison formation of the Colorado Plateau. U.S. Geol. Survey, Bull. No. 1150.

15. KENNARD, T. G., and D. H. HOWELL, 1941: Types of coloring in minerals. The American Mineralogist, **26** (7), 405—421.

16. KIESLINGER, A., 1949: Der Stein von St. Stephan. Wien: Herold Verlag, 486 p.

17. KIESLINGER, A., 1964: Die nutzbaren Gesteine Salzburgs. Salzburg-Stuttgart: Das Bergland Buch, 436 p.

18. Kodak Notes on Practical Densitometry. Kodak Sales Service Pamphlet No. 59 (no date given).

19. MUNSELL, A. H., 1947: A color notation (18th edition, revised). Baltimore, Md.: Munsell Color Company, Inc., 74 p.

20. MUNSELL Soil Color Charts, 1954: Baltimore, Md.: Munsell Color Company, Inc.

21. NICKERSON Color Fan, maximum chroma, 40 hues. (no date given), Baltimore, Md.: Munsell Color Company, Inc.

22. PATNODE, H. W., 1941: Relation of organic matter to color of sedimentary rocks. Am. Assoc. Petrol. Geol., Bull., **25**, 1921—1933.

23. TOMLINSON, C. W., 1916: The origin of red beds. J. Geology, **24**, 153—179.

24. WINKLER, E. M., 1966: Memorial Library, University of Notre Dame Stone Mosaic. Earth Science, **1966** (2), 56—58.

25. WRIGHT, W. D., 1969: The measurement of colour. London: Adam Hilger, Ltd., 340 p.

5. Decay of Stone

5.1. Prologue to Weathering

The accelerating rate of decay of cultural treasures of stone and concrete is becoming a familiar story. Fig. 72a, 72b, 72c suggest the progress of decay in a sandstone sculpture, exposed to the elements since 1702 and photographed in 1908

Fig. 72. Stone decay in the industrial atmosphere of the Rhein-Ruhr; sculpture is of porous Baumberg sandstone (Upper Cretaceous) at Herten Castle near Recklinghausen, Westphalia, Germany, built in 1702. a Appearance in 1908, showing light to moderate damage, b appearance in 1969, showing almost complete destruction and c estimate of the change in rate of decay since 1702. Photos and information supplied by Dr.SCHMIDT-THOMSEN, Landesdenkmalamt, Westfalen-Lippe, Muenster, Germany

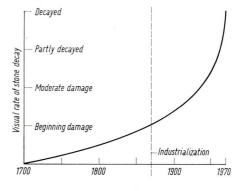

and 1969. The weathering damage in the first 200 years was relatively mild compared with that suffered in the 60 years of the present century.

The process of weathering is merely the adjustment, or readjustment of minerals and rocks to conditions prevailing at the earth's surface, by conversion of existing minerals to minerals of higher stability towards the atmosphere. Many rocks were formed under much higher temperatures and pressures than now seen at the earth's surface. The presence of oxygen leads to oxidation, the presence of moisture to hydration or to solution. The readjustment may occasion volumetric expansion of the crystal lattices. Depending on the "tightness" of the crystal

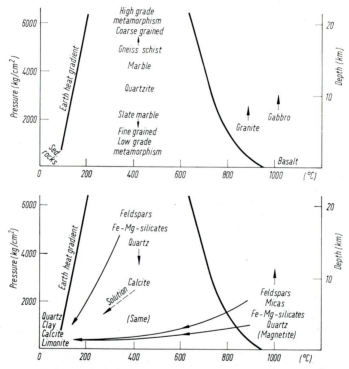

Fig. 73. Environment of formation of minerals and rocks and paths of mineral weathering

lattice the process of readjustment may be slow as a geological process; it may be, however, rapid enough to inflict damage in less than a generation. The urban atmosphere of the twentieth century creates special environmental problems to exposed stone surfaces and can accelerate the process of weathering to many times that of natural rural environs. The charts of Fig. 73 present the original environment during the formation of minerals and rocks. Arrows mark the direction from the original mineral towards the more stable weathering product. Feldspars and the ferromagnesian silicates hydrate to clays, but the micas — after rapid loss of the iron — and quartz are very stable and remain in their original form for thousands of years or more. Carbonate and sulfate minerals dissolve.

The chemical process of weathering is aided by mechanical break-up which leads to the rapid enlargement of the mineral surfaces, providing more ready accessibility to oxygen and moisture, and further and accelerated chemical destruction. The degree of destruction depends on the climate and local circumstances. The chapter on weathering agents offers some detailed information on the corrosive urban environments conducive to stone decay. Soil — the weathering end product — is not discussed here, nor the interaction of soils with stone.

5.2. Weathering Agents

The decay of stone and concrete in engineering structures and monuments is closely related to the geologic process of rock weathering; most of the decay progresses above the ground sufrace, as it is influenced by the following important weathering agents: atmosphere, rainwater, running water, lake waters, seawater. The atmosphere and rainwater are most instrumental in the decay of stone. The influence of plants, bacteria and animals on stone decay is discussed in a separate chapter.

Atmosphere

The atmosphere surrounds us and supports plants and animal life with a basic composition of 78% nitrogen by volume, 21% oxygen and 1% CO_2, argon, and other gases. The atmosphere also supports a reservoir of aggressive impurities such as H_2O, SO_2, SO_3, NO_2, Cl_2, etc. These eventually settle out on the stone surface as "aerosols", the dispersed state of matter in a gaseous medium, and react with stone in aqueous solution. Aerosols comprise sizes ranging from molecules to raindrops. Fig. 74 gives the size distribution of aerosols. The size range

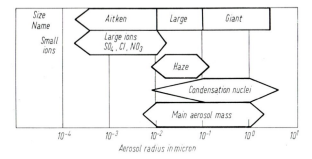

Fig. 74. Sizes and classification of aerosol particles. After JUNGE (1958)

of 10^{-1} to 10^{-3} microns appears to be most important in the decay of stone: these are the larger sized ions, such as the corrosive sulfate and others. Some of the ingredients, e.g, the sulfate and the chloride, may be added from vast natural sources, the oceans and salt flats. Continuous mixing of the air masses by winds and updrafts on the one hand, and the partial neutralization of acids by dust in suspension on the other prevent the concentration of pollutant ions in the air to beyond human tolerance. According to FENN et al. (1963), industrial pollutants have not yet penetrated the global atmosphere; most of the pollutant ions have returned to the ground quite rapidly or were neutralized faster than we realize. The intensity of air pollution is reflected in the quantity of suspended matter which settles as soot or dust, or remains in suspension as aerosols. The number of small-sized particles of 10^{-3} to 10^{-1} microns (Aitken size) contrasts polluted, dust-laden city atmospheres with clean air over oceans and on high mountains (Table 19). The number and size of particles decreases with height above the ground but increases with rising relative humidity.

The chemical attack on stone is due largely to the solvent action of water and its dissolved impurities, including carbon dioxide, inflicting acid corrosion.

Table 19. *Concentration of Ion-sized Particles in the Atmosphere*

Location	Av. Number per m.^{3} air
City	147,000
Town	34,000
Country — inland and seashore	9,500
Mountains: 500—1000 m.	6,000
1000—2000 m.	2,130
above 2000 m.	950
Islands	9,200
Oceans	940

After: JUNGE (1958).

Carbon Dioxide

Carbon dioxide is a minor constituent of the atmosphere with an average of 0.034% (about 306—318 ppm), but the figure was somewhat less before the major global industrialization had started. The concentration increases in cities to as much as 0.27%. The CO_2 increase in the atmosphere may be traced to industrial smoke, automotive exhaust and increased biospheric activity. The rate of increase in the atmosphere was estimated to be 0.1 to 0.3% a year (BOLIN and ERICKSON, 1959). On this basis, Fig. 75 presents the total CO_2 output through combustion of fossil fuels in the upper curve, the range of the actual CO_2 increase in the atmosphere in lower curves. The strong increase seems to stem from two major sources, biospheric and industrial.

a) *Biospheric cycle:* The biospheric cycle is a part of the organic disintegration; the photosynthesis of plant life consumes large quantities of CO_2 during daylight hours when oxygen is produced; this process reverses during the night hours. The values reach a maximum early in the morning and decreases during the day as part of the diurnal cycle. The concentration near the ground surface may fluctuate between 200 and 600 ppm from soil and vegetation. Some CO_2 increase may be traced to the conversion of semi-active grassland and prairie to more active growing crops often supported by extensive irrigation.

b) *Industrial sources:* Industrial sources are a major supply for CO_2 in the atmosphere. Automotive exhaust is credited with more than 60% of all industrial exhausts; the caloric efficiency depends mostly on the fuel-to-air ratio: rich engine mixtures are prevalent in city traffic with frequent stops at numerous intersections. An idling engine emits much CO and unburned hydrocarbons

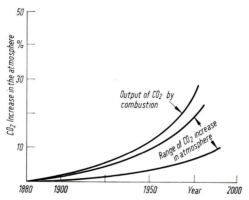

Fig. 75. World-wide increase of the CO_2 content in the atmosphere due to fossil fuel combustion. From BOLIN and ERICKSON (1959)

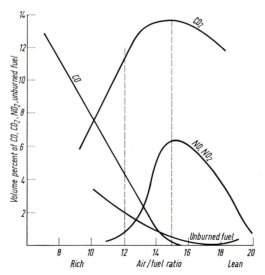

Fig. 76. Amount of CO, CO_2, NO and NO_2 in the automotive exhaust. Output compares rich fuel mixture with lean fuel mixture. After BOLT (1966)

(Fig. 76); complete combustion eliminates these but increases CO_2 and NO_2. The sunshine in a closed-atmosperic basin can convert unburned hydrocarbons to photochemical smog; the photochemical smog of the Los Angeles Basin is a much cited example. The nitrogen of the hydrocarbons oxidizes to NO_2 and to corrosive nitric acid under the influence of strong oxidants. The compulsory installation of afterburners is expected to provide complete combustion in slow city traffic, eliminating toxic CO and unburned hydrocarbons; this would, however, raise the present CO_2 output of $10-12\% \ CO_2$ in the exhaust to almost 14%. The CO_2 level in the atmosphere will probably continue to increase unless we shift to entirely new principles of automative power or different means of transport, such as all-electric rapid transit, with nuclear generation of electricity.

Carbon Monoxide

Incomplete combustion emits millions of tons of carbon monoxide annually. CO is very toxic to both man and animal: rapid depletion of the oxygen level in the blood leads to unconsciousness and death. CO is not corrosive to stone nor does it oxidize to CO_2 in nature, but acts as a catalyst in the oxidation of SO_2 to SO_3. There is evidence that CO disappears into outer space or is oxidized by soil bacterial action.

Sulfates, SO_2, SO_3

The combustion of the fossil fuels coal, oil and natural gas releases sulfur as SO_2 into the atmosphere. Coal is the greatest offender with a sulfur content of as much as 8%, natural gas the least offender with a maximum of 2%. Major cities

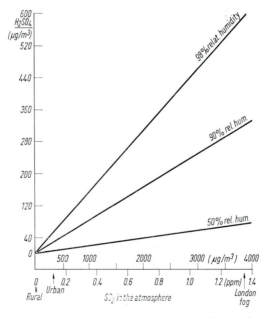

Fig. 77. Conversion of SO_2 to sulfuric acid mist in the atmosphere, induced by humidity as a catalyst. Anonymous (1959)

restrict combustion to low-sulfur coal with less than 1.5% total sulfur. Industrialized countries should generally assume a consumption of one ton of coal a person per year (including industry). This 0.6 kg. of concentrated sulfuric acid per person per year would be produced at an average of 50% relative humidity of the atmosphere, but 4.5 kg. of sulfuric acid at 98% relative humidity, and still more in fog with near 100% relative humidity (data from Fig. 77). High population density in cities can produce, therefore, considerable quantities of sulfuric acid in the atmosphere. The deleterious effect of such acid concentrations on life and materials is well known.

Roasting of sulfide ores in smelters spews out huge quantities of SO_2 through high smoke stacks. Many square miles of barren ground near such smelters offers eerie witness of excessive sulfate fallout. The area of Ducktown, Tennessee, Sudbury, Ontario, and others are well-known examples. The minerals pyrite, marcasite and gypsum, as well as organic sulfur in fine distribution are responsible for the sulfate emission from coal. Some of the sulfur in the atmosphere, however, finds its source in ocean spray and dust from desert flats. In rural areas sulfur may also derive from the oxidation of H_2S rising from the soil in small quantities in summer.

Table 20. *Sulfate Emission for the Years 1963 and 1966, by Source*

Source	% of SO_2	
	1963	1966
Burning of coal:		
power generation	41.0	41.6
other combustion	19.0	16.6
Combustion of petroleum products	20.7	19.6
Refinery operations	6.8	5.5
Smelting of sulfide ores	7.4	12.2
Coke processing	2.0	1.8
Sulfuric acid manufacture	1.9	1.9
Coal refuse banks	0.8	0.4
Refuse incineration	0.4	0.4

From: Anonymous, U.S. Dept. of Health (1969).

The heating of homes and offices in winter raises the sulfate output, often to high levels. SO_2 can slowly oxidize to the much more dangerous SO_3 at a rate of about 0.1 to 0.2% an hour in the atmosphere under conditions of intense sunlight, but much faster in the presence of catalytic oxidants, such as high relative humidity of the atmosphere and the presence of CO (see Fig. 77). Table 20 gives an analysis of the sources for sulfate emission in American cities:

SO_2 measurements in various atmospheres can be given as one-hour maxima, 24-hour maxima, or mean annual. Table 21 gives the SO_2 content of various atmospheres. The data should be accepted with a certain caution, because different recording techniques were used by different agencies and countries.

The sulfate attack on carbonate rocks proceeds in a dual way, as solution by the action of sulfuric or sulfurous acid, and by the conversion of carbonates to either calcium sulfite or calcium sulfate. Calcium sulfite has approximately the solubility of calcite in pure water but is neither influenced by temperature nor by the CO_2 content of the solvent. Calcium sulfate is much more soluble than both calcite and calcium sulfite. The sulfates of magnesium are even more soluble than gypsum.

The effect of sulfate attack on silicate rocks is barely known but probably enhances the process of kaolinization. Aerobic sulfate bacteria, no doubt, participate in the processes.

Table 21. *Sulfate Content of Various Atmospheres*

Location	micrograms/m.^{3} (μg/m.^{3})		
Rural England	55		
London (mean conc.)	285		
London (St. Pancras station) summer	100		
winter	500		
London, killing fog of 1952, winter	3840		
Pulp mill, surrounding (New Hampshire)	5720—37,200		
New York City, annual average	486		
New York City, 24 hours average	1085		
Kansas City, Mo., annual average	5.7		
Chicago, annual average	286—542		
Denver, Colo.	28.6		
Rhein-Ruhr-Industrial area (ann. mean)		(1963/64)	(1967/68)
Gladbeck		290	180
Essen		270	140
Recklinghausen		210	170

Data from: Air quality criteria for sulfur oxides (1969).
         Reinhaltung der Luft in Nordrhein-Westfalen (1969).

Chloride

Chloride is an important constituent of the atmosphere. The sources are marine, desert, and industrial; salt dusts from dried-up playa lakes and salt spray play a subordinate role. Some of the chloride may convert to hydrochloric acid which readily dissolves carbonate rocks. Metal corrosion near ocean shores is conspicuous.

Nitrates

Nitrates, usually as NO_2 are uniformly distributed throughout the atmosphere, about 2 to 4 micrograms per cubic meter, but registers much higher concentrations in urban and industrial areas (up to 400 μg/m.^{3} in Los Angeles). Complete automotive combustion is an important contributor of nitrogen oxides to the atmosphere. Conversion to corrosive nitric acid is common in photochemical smog.

5.3. Removal of Ingredients from the Atmosphere

The present strong concentration of atmospheric ingredients from air pollution would have made our atmosphere unlivable if the removal of such ingredients did not take place continuously; this keeps the level of toxic substances still far below the upper limit of tolerance in spite of local higher concentration within and around industrial and human centers. JUNGE (1958) considers several possibilities of natural removal:

1. Removal of aerosols:
   a) fallout due to gravity as dust or soot, the "dry fallout";
   b) impact and capture of particles on obstacles at the earth's surface;
   c) washing out by precipitation.

2. Removal of gases:
   a) escape into space of light CO, H_2, N_2O, some CO_2, CH_4;
   b) absorption, decomposition or both at the earth's surface by vegetation and mineral weathering or hydrolysis;
   c) decomposition in the atmosphere by reactions resulting in the formation of aerosols or other gases (formation of smog);
   d) washout by precipitation.

Washout by precipitation appears to be the most important and most complex process.

5.4. Rainwater in Rural and Polluted Areas

The composition of rainwater is closely associated with the composition of the atmosphere; the strong corrosive action is based on the ions which are picked up during the travel through the atmosphere. WHITEHEAD and FETH (1964) distinguish rain, dry fallout and bulk precipitation which is the sum of both. The ions of rainwater show the marine effect and the composition of the atmosphere whereas the dry fallout reflects mainly water-soluble contributions from local dust sources between rainfalls. CARROLL (1962) discusses the hardness and ion content

Table 22. *Chemical Rainwater Analyses, in* ppm

Location	SO_4	Cl	Ca	Mg	Na	NO_3	pH
Chamberlin glacier, Alaska (1) (100% rural)	0.4	0.0	0.7	0.1	0.3	—	—
Northern Europe (2)	2.19	3.5	1.4	0.4	2.1	0.3	5.47
Chicago, O'Hare Airport, suburban (3)	8.1	1.5	2.9	—	0.7	1.0	5.96
Knoxville, Tenn. (urban) (2)	14.5	6.5	2.8	0.7	3.6	—	—
Knoxville, Tenn. (suburban) 5 km. from downtown (2)	5.6	3.3	0.9	0.4	1.5	—	—

From: (1) after FREY (1963); (2) after GORHAM (1961); (3) U.S. Public Health Service (1962).

of rainwater and its influence on the chemical composition of surface waters. In areas of high carbonic and sulfuric acid in the air, rain may become very acid and is then extremely corrosive to both stone and metals. Rainwater analyses from very different natural environments are given in Table 22.

The sulfate data of Knoxville, Tenn. for 1927 are probably no longer valid but the urban-suburbarn contrast is significant.

The washout rate of aerosols from the atmosphere by precipitation depends on the raindrop size, the drop speed, and the distance between the cloud base and the land surface. The absorption equilibrium of raindrops is reached much faster with slowly falling small droplets of larger specific surface than with large drops of greater falling speed in semi-arid and arid areas during infrequent driving rains.

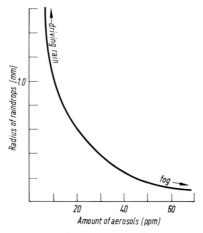

Fig. 78. Maximum absorption of aerosols in rain. Raindrop size versus absorption of aerosols. After JUNGE and WERBY (1958)

Fig. 78 presents the relation of rain-absorbed aerosols to the radius of the rain-drops: Large rain drops can absorb a maximum of 10 ppm whereas fog droplets 60 ppm or more. Urban winter fog can therefore be very corrosive and toxic, as it was in the London fog of 1952 which killed over 4000 people.

Carbon Dioxide in Rainwater

Cold water near freezing can dissolve twice as much CO_2 as water near 25° C. NORDELL's (1951) CO_2 equilibria with rainwater are plotted in Fig. 79. The abscissa gives the CO_2 content of the atmosphere for rural, urban, and urban industrial atmospheric CO_2 levels, marked on the graph with arrows, the ordinate gives the CO_2 solubility in water. The quantity of dissolved CO_2 is essential in that it influences the solubility of carbonates and accelerates the decomposition of silicate rocks. Rain may not have reached equilibrium if the time of interaction with the atmosphere was too short. The pH of rainwater lowers as the CO_2 in solution rises.

Sulfate and Chloride in Rainwater

Sulfate is the strongest of the corrosive pollutants in rain as it may in part hydrolize either to sulfurous acid from unoxidized SO_2 or to the much more corrosive sulfuric acid from SO_3. The supply of chloride and sulfate is extensive near the sea shores and over deserts, as well as over industrial areas such as the Chicago-Detroit-Cleveland area and others: Fig. 80 shows the increase of chloride and sulfate in the waters of the Great Lakes through progressive industrialization: chloride and sulfate almost doubled in the last 50 years in Lakes Erie, Ontario, and Michigan. Na, K, and Ca also increased at the same time. The figures appear to be good evidence that weathering rates of soils and rocks in contact with the lake waters are doubled (WINKLER, 1970). In the Great Lakes

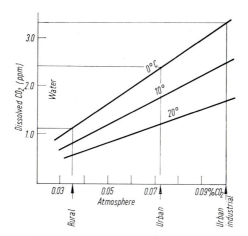

Fig. 79. Saturation equilibrium of carbon dioxide in the atmosphere with water, as dissolved CO_2. After WINKLER (1966)

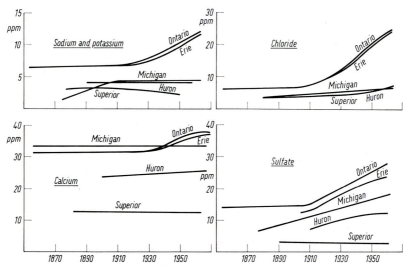

Fig. 80. Increase of the ion content, pollutants and nonpollutants in the waters of the Great Lakes in the last 50 years. From WINKLER (1970) after BEETON (1965)

region, Na and K were obviously removed from the igneous and metamorphic components of the glacial drift of the polluted lakes Michigan, Erie, and Ontario, whereas Ca increases in Lake Erie and Ontario where more carbonate rocks are exposed along the shore lines. The rate of ionic increase in the Great Lakes should be a stern warning, of what the damage to stone and concrete by polluted waters may be expected to be in the future. Shoreline fortifications, built with dimension stone, rip-rap, or concrete, should therefore be well selected in polluted waters, as they are continuously pounded by wave action or washed by rapidly passing corrosive stream waters.

5.5. Dusts as Neutralizing Agents of Acid Rains

Natural dust, in contrast to dust from other sources, is particulate matter ranging from about 1 to 100 microns in size, mostly composed of terrestrial wind-born debris, some volcanic dust, and minute amounts of extraterrestrial cosmic dust. Dust suspended in the lower atmosphere is often washed out by falling rain drops as muddy rain; some dust settles between rainfalls as "dry fallout". The velocity of fall is determined by the grain size.

a) *Windblown dust:* Silt-sized particles may be picked up by high winds from dry river beds and flood plains during low-water stages and beaches. More frequently, however, dust pickup is facilitated from plowed fields and construction sites lacking a protective vegetation cover. Despite local accumulation of wind-blown debris in the geological past, the rate of dust deposition is believed to have at least tripled since man has started agricultural activity by scarring the protective sod of the American prairies. The calcite component of the dust readily reacts with the acids in the rainwater despite the brief contact between the cloud base and the surface of the terrain. Calcite does not eliminate the sulfate ion from the rain but rather converts it from sulfurous or sulfuric acid to the relatively harmless calcium sulfate, gypsum. The sulfate ion tied up in gypsum is not noxious to man; waters rich in sulfate, however, are known to have severely attacked alkalic cement in concrete (KIESLINGER, 1962). The water-soluble gypsum appears to explain the relatively high calcium content of rainwaters (SUMI *et al.*, 1959). Concerning the acidity of rain in the future, the question may be raised whether or not man's interference in two opposite directions will reestablish original pre-interference condition in the composition of the rainwaters, by higher acidity of the rainwater through pollution on the one hand, and higher production of neutralizing dust on the other.

b) *Volcanic dust:* The explosive eruption of volcanoes has sent dust, essentially silicate glass into the upper atmosphere, where it has travelled for several years before it could settle back to the earth's surface. Not much reaction should be expected between volcanic dust and rainwater.

c) *Extraterrestrial dust:* Cosmic dusts are found in marine sediments of most geological ages as occasional small pellets of iron or silicate glass. Their chemically inactive characteristics and the minute quantities present make their effect insignificant.

The degree of washing of polluted air by rain is very difficult to assess and the implications are difficult to foresee. MEADE (1969) calculated the atmospheric contribution to the dissolved loads of streams (excluding bicarbonate), 20% in North Carolina and 50% in New Hampshire after reaction of the rain water with the natural dusts. The abundance of ions in streams and lakes appears to stem from a combination of sources: polluted waters, washout from polluted air, and higher water activity by the increase of CO_2 and SO_4.

5.6. Stone Weathering Compared with Atmospheric Metal Corrosion

The corrosiveness to metals of different atmospheric environments is well known to the corrosion engineer. He distinguishes crudely the corrosive environments as dry inland, rural unpolluted, semirural, marine, industrial marine, and industrial. It seems likely that the weathering rates of ferrous-ferric oxides and hydrates as well as of ferrous-ferric silicate minerals may approach the corrosion rates of iron. Conductive minerals, such as pyrite, marcasite, hematite, etc., may also act as galvanic couples if aided by an electrolyte and may be subject to electrochemical corrosion attack. Since conducting minerals are similar to metals, they also should be subject to electrochemical attack by differential aeration. If two electrodes of the same metal are in a dilute NaCl solution and the electrolyte around one electrode is aerated, this electrode becomes the cathode and is protected while the other electrode in contact with a non-aerated electrolyte becomes the anode and is attacked. The differences in oxygen concentration produces a potential difference and causes current to flow. This type of cell causes severe attack to occur in crevices and in areas shielded from oxygen by corrosion products or debris. The disintegration of granite located next to marble is often attributed to the reaction of dissimilar stones (like dissimilar metals) with one another. A reaction between the two different rock types is very difficult to visualize; dissolved components, however, may infiltrate from the soluble marble and enter the granite, leading to salt action (see chapter Moisture and Salt Action).

Table 23. *Corrosion Rates of Iron in Different Atmospheres* (open-hearth iron specimens, $2 \times 4 \times \frac{1}{8}$ inch)

Location	Type of Atmosphere	Av. Weight loss, g./year	Relative Corrosiveness
Karthoum, Egypt	dry inland	0.16	1
Abisco, North Sweden	humid, unpolluted	0.46	3
Singapore, Malaysia	tropical marine	1.36	9
Daytona Beach, Fla.	rural subtropical	1.62	11
State College, Pa.	rural, moderate humid	3.75	25
Miraflores, Canal Zone	tropical marine	4.5	31
Kure Beach, N.C. (800 ft. from ocean)	marine	5.78	38
Sandy Hook, N.J.	marine, semi-industrial	7.34	50
Pittsburgh, Pa.	strongly industrial	9.65	65
Frodingham, British Isles	industrial	14.81	100
Daytona Beach, Fla.	marine subtropical	20.43	138
Kure Beach, N.C. (80 ft. from ocean)	marine	70.49	475

From: LARRABEE and MATHAY (1963).

LARRABEE and MATHAY (1963) compare different corrosion rates of openhearth iron specimens which were exposed to a variety of atmospheres; the figures give an indication of the aggressiveness for different relative atmospheres. Table 23 compares the values.

What remedy can be taken to preserve or improve our present atmosphere, and rainwaters? The power industry appears to develop in the right direction: accelerated change-over to safer atomic power, more efficient electric power and possibly to solar power where possible; these changes should at least stabilize the present acute problem of direct air pollution and indirect water pollution. The greatest present offender, automotive transportation, is expected to change over to new fuels and combustion systems. Much research must still be done to achieve this goal.

Summary

The important agents of weathering are the atmosphere and rainwater. CO_2 and SO_4 are the most important deleterious ingredients; both are emitted by the combustion of fossil fuels.

The natural removal of these ingredients from the atmosphere takes place as dry fallout, washout by rain, reaction of acid pollutants with inorganic suspended dust, and possibly escape into outer space. Construction activity is freeing more dust to encounter more man-made pollutants.

Stone corrosion is compared with metal corrosion in atmospheres of different corrosivities. The increased output of pollutants has stepped up the destruction of stone monuments many times since the begin of industrialization in the last 100 years. The time-lapse photos of the destruction of monuments near the highly industrial Rhine-Ruhr area do not require further comment (Fig. 72 a—c).

References

1. Anonymous, 1962: Air pollution measurements of the National Air Sampling Network. U.S. Public Health Service, Public Health Report 1957—1961.
2. Anonymous, 1969: Air quality criteria for sulfur oxides. Consumer Protection and Environmental Health Service, Natl. Air Pollution Control Administration. U.S. Dept. Health, Education and Welfare, Public Health Service, Washington, D.C.
3. Anonymous, 1969: Reinhaltung der Luft in Nordrhein-Westfalen. Rept. Conference on Air Conservation in Düsseldorf, Oct. 13—17, 1969. Ministry of Labor and Welfare, Nordrhein-Westfalen, Germany.
4. BEETON, A. M., 1965: Eutrophication of the St. Lawrence Great Lakes. Limnology and Oceanography, **10** (2), 240—254.
5. BEETON, A. M., and D. C. CHANDLER, 1963: St. Lawrence Great Lakes. In: Limnology in North America (D. G. FREY, ed.). Madison, Wisconsin: Univ. Wisconsin Press, pp. 535 to 558.
6. BOLT, J. A., 1966: Future automotive power plants — and air pollution. Proc. Symp. Natl. Conf. Air Pollution, 3rd, Washington, D.C., U.S. Dept. Public Health Service, pp. 85—101.
7. CARROLL, D., 1962: Rainwater as a chemical agent of geologic processes — a review. U.S. Geol. Survey Water Supply Paper 1535-G, 1—18.
8. FENN, R. W., H. F. GERBER, and D. WASSHAUSEN, 1963: Measurements of the sulfur and ammonium component of the arctic aerosol of the Greenland ice cap. J. Atmospheric Science., **20** (5), 466—468.
9. GORHAM, E., 1961: Factors influencing supply of major ions to inland waters with special reference to the atmosphere. Geol. Soc. Am. Bull., **72** (2), 795—860.

10. JUNGE, C. E., 1958: Atmospheric chemistry. Advances in Geophysics (LANDSBERG and MEEHAM, eds.), Academic Press, **4**, 1—108.
11. JUNGE, C. E., and R. T. WERBY, 1958: The concentration of chloride, sodium, potassium, calcium and sulfate in rainwater of the United States. J. Meteorology, **15** (5), 417—425.
12. KIESLINGER, A., 1962: Water tunnels in gypsiferous rocks. Österr. Ingenieur-Zeitschrift, **5**, 338—350.
13. LARRABEE, L. R., and W. L. MATHAY, 1963: Iron and steel. In: Corrosion resistance of metals and alloys (LaQUEE and COPSON, eds.). Reinhold, pp. 305—353.
14. LIVINGSTONE, D. A., 1963: Alaska, Yukon, Northwest Territories, and Greenland. In: Limnology in North America (D. G. FREY, ed.). Madison, Wisconsin: University Wisconsin Press, pp. 535—558.
15. MEADE, R. H., 1969: Errors in using modern stream load data to estimate natural rates of denudation. Geol. Soc. Am. Bull., **80** (7), 1265—1274.
16. NORDELL, E., 1951: Water treatment for industrial and other uses. New York: Reinhold, 526 p.
17. SUMI, L., A. CORKERY, and J. L. MONKMAN, 1959: Ca-sulfate content of urban air. Am. Geophys. Union, Geophys. Monograph # **3**, 69—80.
18. WHITEHEAD, H. C., and J. H. FETH, 1964: Chemical composition of rain, dry fallout, and bulk precipitation at Menlo Park, California, 1957/58. J. Geophysical Research, **69** (**16**), 3319—3331.
19. WINKLER, E. M., 1970: Errors in using stream-load data to estimate natural rates of denudation: Discussion. Geol. Soc. Am. Bull., **81** (3), 983—984.

General References on Rock Weathering

1. KELLER, W. D., 1957: The principles of chemical weathering. Columbia, Missouri: Lucas Brothers, 111 p.
2. KIESLINGER, A., 1932: Zerstörungen an Steinbauten, ihre Ursachen und ihre Abwehr. Leipzig-Wien: Deuticke.
3. LOUGHNAN, F., 1969: Chemical weathering of silicate minerals. New York: Elsevier Publishing, 154 p.
4. OLLIERS, C., 1969: Weathering. Edinburgh: Oliver and Boyd, 304 p.
5. REICHE, P., 1950: A survey of weathering processes and products. New Mexico University Publication, Geology 3.
6. SCHAFFER, R. J., 1932: The weathering of natural building stones. Dept. Sci. Ind. Res., Bldg. Res. Spec. Rept. 18, pp. 1—149.
7. SHORE, B. C. G., 1957: Stones of Britain. London: Leonard Hill Ltd., 302 p.
8. VILLWOCK, R., 1966: Industriegesteinskunde. Offenbach/Main: Stein-Verlag, 279 p.

6. Moisture and Salts in Stone

Moisture and salts are the most damaging factors in stone decay. The very complex capillary system of stone and the often strange paths of moisture in buildings and monuments complicate the understanding of the mechanisms, such as solution, salt migration, salt crystallization, salt hydration, the thermal expansion of entrapped salts, frost weathering, and the expansion of pure water substance in rock pores. Many field data have been compiled both from deserts and from structures in urban humid and semi-humid climate; some theoretical considerations had to substitute for observations and experiments where processes are limited to areas located in narrow capillary systems within the stone. The following topics will be discussed: sources of moisture, moisture transfer mechanism in capillary systems of stone and masonry walls, and origin and behaviour of salts in capillaries.

6.1. Sources of Moisture

Most of the moisture present in stone walls and monuments may be derived from the atmosphere, from rain, and from rising ground moisture. All of these sources are equally important and can be equally dangerous.

a) *Moisture from the atmosphere:* Condensation of moisture on monuments and stone walls by strong temperature gradients may deposit considerable quantities of moisture from air of relatively high humidity against a cold stone wall; this important source is often overlooked. Fig. 81 plots the constant-saturation mixing ratio of moisture in grams per cubicmeter of dry air at sea level; as cooling occurs moisture is freed and condenses; there the journey may begin towards the inside of the stone substance. The amount of moisture freed may be readily seen in Fig. 81. If, for instance, warm air cools at 35° C and 60% relative humidity at point B and shifts at constant moisture content towards 20° C at B′, the relative humidity rises with dropping temperature and starts condensation as soon as the point of 100% relative humidity, the dew point, is passed. At 20° C about 8.5 g. of moisture falls out per cubic meter of air, but 21 g. of moisture if cooling occurs from 40% relative humidity at 40° C to 15° C, from A to A′. The second case may be readily expected to be found both in deserts and on stone walls in cities. Complications enter the picture if we consider moisture diffusion in the air, air currents and other factors. The theoretical curve is merely a model and a point of departure for further calculations.

b) *Moisture from rain:* Driving rain will not enter a stone wall as easily as is generally believed. The duration of rain and fog is too short to permit much moisture to infiltrate; subsequent drying conditions pull out again most of the moisture following a drenching rain. Well-exposed surfaces of permeable sandstone, however, transmit moisture from rain and fog easily and may sometimes permit deep penetration (KAISER, 1929).

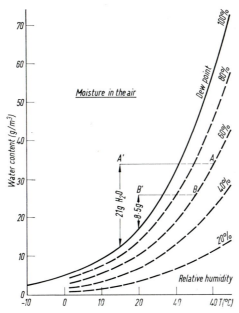

Fig. 81. Water content of air in g/m³ at 1000 mb pressure. Release of moisture is shown during cooling at constant moisture content: in A—A' cooling occurs from 40 to 15° C along a wall resulting in 21 g moisture freed, in B—B' only 8.5 g is freed if air is cooled from 35 to 20° C. Data from Pseudo-Adiabatic Chart, U.S. Weather Bureau, WB Form 770—9

c) *Ground moisture* stems both from splashing rain and from groundwater. Both contain more ions than surface waters as salts on the streets and slow circulation of groundwater between mineral particles of rocks and soils have permitted pickup of ions towards the point of saturation. The upward motion of ions is a continuous process; the water moves towards the open ends of the channelways where the salts may crystallize near or at the stone surface, or, if soluble enough, they may travel back again. The chapter on Weathering Agents provides more information on the ion supply of natural waters.

d) *Leaking pipes and gutters* are often hidden behind mortar and may remain undetected unless dark wet spots are visible on the wall. Condensation of moisture along plumbing and gutters may sometimes be supplying as much moisture as true leakage.

e) *Kitchens and bathroom showers* provide a steady source of moisture if these are not properly ventilated from where travel may start towards the surface.

f) *Construction moisture* may occur if the ventilation of interior parts of a building is insufficient during construction until mortar and plaster have hardened.

Moisture problems in buildings and monuments are usually extremely complex. Older buildings are rarely insulated against rising ground moisture. Faulty insulation in modern buildings may be as bad as none at all.

6.2. Capillary Rise of Moisture

Rise of moisture by capillarity exposes masonry walls and unprotected monuments to ground moisture and sometimes to salt impregnation which is a relatively slow, but cumulative process. The maximum height of capillary rise is often marked by a white efflorescent rim or a dark, wet margin which remains wet as the salt concentration at this level tends to hold the moisture for a long time, especially in very fine, porous masonry, stone, concrete and mortar. KIESLINGER (1957) discusses fluid travel in masonry walls where he observes good correlation between field data and the simple equation for capillary rise height, h,

$$h = \frac{2\,s}{R\,d}$$

s is a constant, about 0.074 g./cm.;
d is the density of water, about 1 g./cm.^{3};
R is the radius of the capillary.

Based on this equation, grain size ranges of 0.02—0.006 mm. should show a maximum vertical rise of 3 to 10 m.; 0.006—0.002 mm., 10 to 30 m. Vos (1968) conducted laboratory experiments for capillary rise, both in horizontal and in vertical direction. Fig. 82 plots time versus travel distance in 0.004 and 0.002 mm.

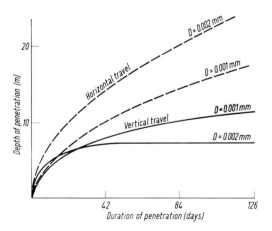

Fig. 82. Capillary travel of water, both horizontal and vertical. From Vos (1968)

channel diameters: water travels twice as far in horizontal direction as in vertical. This surprising result probably explains the often strange horizontal travel routes of moisture in buildings. Vos's data are difficult to compare with the data of KIESLINGER as neither author specifies the material nor is it easy to transpose grain sizes to pore diameters. Wind pressures or rapid changes in barometric pressure are not considered, nor "negative" suction induced by solar radiation and dry winds. Additional complications may arise in the concealment of open ends of capillaries with lichens and mosses. Such an organic cover retains moisture and does not permit evaporation from the open ends of channel ways, so that the wall remains moist.

Capillary water never leaves the capillaries as free flow nor does all the rising moisture reach the maximum height marked by discoloring or efflorescent margins. The distribution of vertical moisture compares well with the distribution of

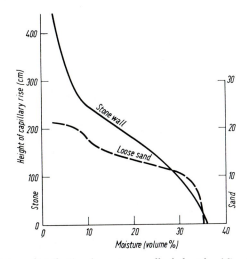

Fig. 83. Vertical moisture distribution in a stone wall of church of San Sebastiano in Venice, compared with moisture distribution in pure sand. From Vos (1968)

moisture in the ground from the surface of the water table upward towards the land surface, the zone of aeration. Fig. 83 measures the vertical moisture distribution in a wall of the church of San Sebastiano in Venice, Italy. The shape of this curve is plotted against a moisture distribution curve in sand after suction. According to both curves the capillary rise of most of the water only reaches half the total height, then rapidly decreases and only 10% of the total water reaches the top of the margin.

6.3. Moisture Distribution in Capillary Systems

Pore systems may be filled with fluid, or may be only filled in part, in which case the moisture continuity is interrupted by air pockets or "islands"; the pore system may also be entirely empty, except for a very thin vapor or sorbed coating along the walls measuring only a few molecules in thickness. Fig. 84 sketches a capillary system with open channels, interconnected large pores and isolated vugs. Transport of solution and salt depends much on the moisture travel. Moisture in stone capillaries thus should be expected to travel as follows:

a) *Capillary travel* is very difficult to see and to assess because the size and size distribution of pore systems is little accessible to observation. In general, sandstones have a more consistent capillary system than sedimentary carbonate rocks. A bubble or water island in a horizontal capillary tube should stand still at the same temperature and pressure prevailing at both ends of the tube. In a conically narrowing tube the water moves towards the narrow end. In the presence of a difference of temperature from the outside of a wall towards the inside,

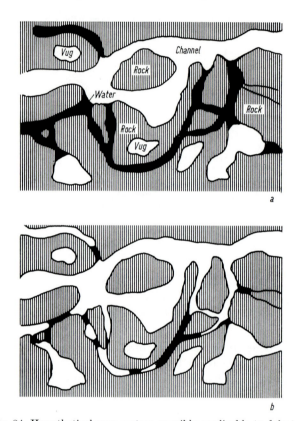

Fig. 84. Hypothetical pore system, possibly applicable to dolostone

A   Only narrower channels are filled with water; thus continuous flow of water is possible through the pore system

B   Water bubbles fill only part of narrow passages between bubbles of vapor. No water motion is possible through the presence of many vapor locks

moisture migrates towards the colder side. This explains the travel of moisture through walls of older buildings from the outside towards the colder interior of the structure.

b) *Capillary condensation* is much influenced by the relative humidity of the air which contacts stone. The maximum hygroscopic moisture content is reached at 98% relative humidity, at which condition all micro-capillaries are filled with moisture as far as they are accessible to water. Capillary condensation also takes place at lower relative humidities to as low as about 50%. Fig. 85 presents the moisture content of stone as it is influenced by the relative humidity of the atmosphere. Vos (1969) experimented with cellular concrete which may compare with a porous sandstone or conglomerate. Soluble salts entrapped in stone increase the hygroscopic moisture to many times the amount in the pure material.

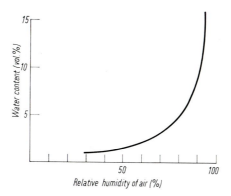

Fig. 85. Moisture content in cellular concrete in dependence on the relative humidity of the air. From Vos (1969)

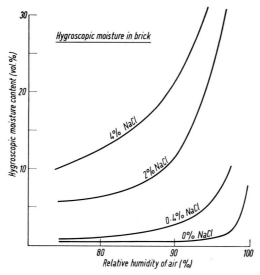

Fig. 86. Hygroscopic moisture in brick depending from the relative humidity of the air and the content of NaCl. From Vos (1969)

Vos (1969) exposed to different humidities brick which were pre-saturated with different amounts of salt (NaCl); Fig. 86 shows the results. These laboratory experiments apply well to stone which is infiltrated with salts near the surface: the presence of moisture absorbed from the atmosphere may be very high, even at relatively low humidities. The soluble salt content of a cross-sectioned sandstone of the Regensburg Cathedral (Fig. 87) could well serve as an example here for the presence and distribution of the salts relative to the stone surface.

Ordered water is strongly held along the rock-water interface by electrostatic attraction in a polar oriented manner; this attraction decreases exponentially with increasing distance from the pore wall (RUIZ, 1962). The water is immobilized and its mechanical properties resemble somehow that of a more or less rigid solid.

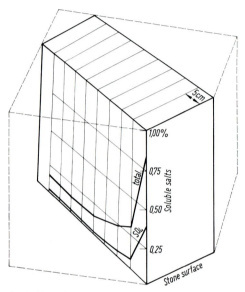

Fig. 87. Distribution of soluble salts across a green sandstone from the Regensburg Cathedral. The corner stone was cut into slices 5 cm thick. The salt migration outward is evident. Adapted from KAISER (1929)

The thickness of this film is about 2 to 3 millimicrons at 50% relative humidity. It is believed that the presence of such ordered water may be instrumental in the known damage done to stone by alternate wetting and drying cycles, osmosis and osmotic pressures during the weathering process, and the presence of non-freezable water. Not all the anomalous behaviour and properties of such water, especially of pore water, should be ascribed to the presence and properties of ordered water.

6.4. Moisture Travel in Walls

Water never creeps upward along the outer stone surfaces; all the water movement occurs within the stone. KAISER (1929) observed a general outward motion of moisture and water-soluble salts with the moisture towards the surface; a cornerstone from the Regensburg Cathedral, Germany, was sectioned into 9 slices 5 cm. thick each, and analyzed for soluble salts. The graphic presentation is given in Fig. 87. The content of calcium, magnesium and sulfate is greatest immediately at the surface, but drops off rapidly a few centimeters away from the surface, to a fraction of the salt content at the surface. Recently, MAMILLAN (1968) recorded a strong increase of sulfate in carbonates, about 5 to 10 times more in the first millimeter beneath the stone surface than further inside. Surface transport of salts in porous rocks from the inside of the stone substance leads to softening and crumbling of the stone combined with a rapid decrease of the general stone strength. KAISER also observed an increase of the rock density at the surface: as much as 20% for soft and friable rock, 5 to 10% for semi-dense rock, only 1—5% for dense rock. All soluble salts tend to move outward. The more soluble chlorides

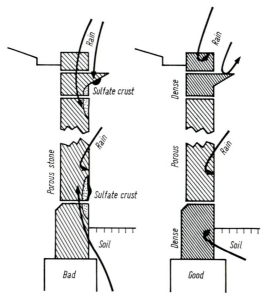

Fig. 88. Travel of moisture in a stone wall, moisture from both precipitation and from the ground. The effect of non-absorbent stone is sketched. From MAMILLAN (1966)

Table 24

Salt	Stone Exposed to Light	Stone Exposed to Shade
Gypsum	2.5%	1.74%
Calcium sulfite	0.41%	0.2%

Data from: KAISER (1929).

and some sulfates remain in solution, and move back and forth with changes in weather and moisture gradient; whenever these salts reach the surface temporarily, driving rains may wash them off and move them back into the street from where a part may enter the structure once again. The less soluble salts crystallize at or near the surface as gypsum, Ca-sulfite, secondary calcite, or as other compounds. KAISER has also observed that the amount of gypsum and Ca-sulfite is greater in light and in parts exposed to the sun than in darker areas. Table 24 compares these figures.

Light-colored areas on stone result from fast and efficient drying, darker colored areas from the inability of moisture to escape properly. Moisture retention is further enhanced by the presence of bacteria, algae and lichens which choke the open ends of the capillaries, thus preventing proper drying. Fig. 88 sketches the travel of moisture in a wall of a building; damage by moisture can be reduced, but not halted by using denser stone near the top of the building as a moisture barrier against rising ground moisture as well as against transmission of rainwater.

6.5. Disruptive Effect of Pure Water Substance

Damage to stone by repeated wetting and drying is well known to the stone industry. DUNN and HUDEC (1966) believe that the disruptive force of water is generated by the expansion of sorbed and ordered water. Based on laboratory experiments with carbonate rocks, those specimens were found to be least sound which

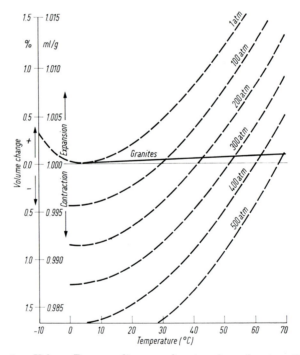

Fig. 89. Temperature-Volume-Pressure diagram of water above freezing. Dashed contours are lines of equal pressure of confined water

recorded a maximum amount of nonfreezable water to a temperature as low as −40° C. Ordinary water substance, in contrast to ordered water, expands sufficiently enough to disrupt stone of low tensile strength when exposed to warming or heating. Fig. 89 plots the volume-temperature relation from data after DORSEY (1940). The abscissa marks the rock temperatures commonly found in stone, the ordinate the volume change of water in millimeters per gram at 4° C as the zero point. Contours connect points of equal volume change for 1 atm, 100 atm, 200 atm, etc. The average volumetric expansion of granite is plotted in the same graph to show the possible pressure which expanding water may exert in granite; granite only expands 0.15% to about 60° C whereas water expands more than 1.5% if uncompressed. If the walls of small pores prevent water from expanding 1.5% at 60° C, water exerts about 380 atm (about 5600 psi) pressure against the pore walls of the granite. Light-colored, sound igneous rock is not believed to reach surface temperatures near 60° C; dark igneous rock, however, may be endangered. At a compressive strength of about 25,000 psi, 7500 psi tensile strength may readily develop and lead to disruption.

Stone disruption by wetting-drying cycles: The disruption of stone by alternate wetting and drying without change of temperatures is long known, and is discussed in detail by HOCKMAN and KESSLER (1950). The authors found rock expansion by moisture for granite ranging from 0.0004—0.009%, with an average of 0.0039%, marbles 0.001—0.0025%, but quartz sandstones 0.01—0.044%. Rapid wetting and drying may be the cause of stone disruption. Ordered water probably aids in the disruption process.

Stone disruption by insolation: The Webster Dictionary defines "insolation" as "the rate of delivery of all direct solar energy per unit of horizontal surface". Stone bursts in the desert are frequently observed by travelers, associated with a loud, explosion-like noise. This phenomenon was ascribed to differential expansion and contraction of minerals in rocks under the influence of high temperature changes in the daily heat cycle. GRIGGS (1936) proved with experiments that the differential expansion of the different minerals in rock could not induce bursts; it is the capillary moisture in the stone which is instrumental in stone decay leading to flaking and bursting. GRIGGS exposed dry granite to 89,400 heating-cooling cycles between 32 and 142° F, which corresponds to 244 years of exposure to temperature extremes. A great change was observed with moist granite after 10 days of experiments which corresponds to only $2\frac{1}{2}$ years of exposure to desert conditions; after this time visible alteration and damage were evident. These experiments suggest that pure water has the disruptive effect which is so often found both in the desert and in urban buildings.

6.6. Transport of Salt in Stone

Natural stone may contain water-soluble salts entrapped in its pores as a natural constituent; more often, however, salts have migrated into the stone after the stone has been set into place; the salt may have come either from the ground, from the street, or from the stone surface through the interaction of polluted air with the stone, from whence salt circulation started in the masonry wall in different directions. Salt can act in stone in several different ways, influenced by the mode of salt travel, temperature and salt concentration.

a) *Ionic diffusion:* Ionic migration of salts can take place as diffusion in water-saturated rock from areas of higher ionic concentration towards areas of lower concentration. The ionic flow is reversed from lower to higher concentrations in osmosis, in which process a semipermeable membrane is to be crossed by the water. The rate of diffusion depends upon both the temperature and the molecular concentration and gradient. GARRELS *et al.* (1949) demonstrated diffusion in water-saturated dense limestone at room temperature and atmospheric pressure; for better visibility of the penetration rate, the graph of Fig. 90 refers to potassium permanganate.

Concentric rings of ferric hydroxide are common in partly weathered silicate rocks and also in sandstones. CARL and AMSTUTZ (1958) showed with the help of laboratory experiments that true diffusion rings may develop during weathering processes as Liesegang Rings. Liesegang's original experiments were made with

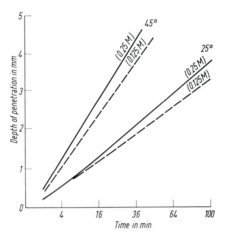

Fig. 90. Rate of diffusion of potassium permanganate at different molar strength through dense limestone, watersaturated. After GARRELS, etc. (1949)

gelatin impregnated with potassium dichromate through which silver nitrate was diffused and precipitated as rings in the gelatin, concentric about the point of injection. Limonite banding is probably formed in weathered rocks by change in the pH of the pore fluid, probably from acid towards more alkalic, leading to the precipitation of very sparingly soluble ferric oxide or hydroxide rings. The color effect of Liesegang Rings is discussed in the chapter on Color and Color Stability.

b) *Capillarity or diffusion:* Rings and bands of ochre-brown iron oxides in sandstones and other weathered rock types are often ascribed to processes of diffusion. The concentric bands usually run parallel to the original stone surface or to other openings in the rock, such as joints or faults. Porous sandstone should transport moisture rather by capillary action than by ionic diffusion, whereby the bands are believed to correspond to the outer fringes of transport of the soluble ferrous iron before it oxidized to insoluble brown ferric hydroxide. If capillary action prevails, the distance of the rings grows narrower outward and finally several rings may join to form one very thick band (LIESEGANG, 1945). The Briar Hill sandstone of Ohio should be cited as an axample (see Fig. 70 in chapter 4). True diffusion develops in a gelatinous medium, such as is found in silica gels of agates. The weathering process of silicate rocks may provide conditions where silica may become mobilized and can form an interstitial gelatinous mass of silica: the ring spacing increases with the distance from the center of supply, provided that the liquid supply remains constant. Banding due to ionic transport by capillarity as well as diffusion is believed to exist in nature.

6.7. Ionic Osmosis and Osmotic Pressures

Osmosis of salts through a solid occurs when a solution is separated from its pure solvent by a semi-permeable membrane, such as like leather. Fig. 91 shows the principle of osmosis: the pure solvent, n_1, attempts to enter and dilute the solution, n_2, through a semi-permeable membrane, to equalize the solutions,

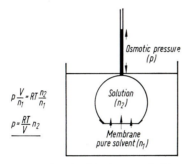

Fig. 91. Equation and sketch of ionic osmosis. From MAHAN (1962)

whereby an osmotic pressure, p, develops against the walls: the magnitude corresponds to the height of the meniscus, multiplied by the density of the solution. VANT' HOFF's equation was modified by MAHAN (1964) to yield:

$$p = \frac{R \cdot T}{V} \cdot n_2$$

In this equation R is the gas constant, 0.082 liter atmospheres per mole degree, T the temperature in degrees Kelvin, V the volume, n_2 the number of moles of solute. Calculations were made by WINKLER and SINGER (in press) for solutions of $MgSO_4$, Na_2SO_4, $NaCl$, and $MgCl_2$. The plots in Fig. 92 give expected osmotic pressures, based on the solute concentration of the salts and on the temperature of the rock substance. Theoretical pressures may be scaled off for various temperatures and solute concentrations, for undersaturated conditions as a full line, for rare conditions of supersaturation dashed. Maximum osmotic pressures are very difficult to evaluate in stone as such pressures are usually superimposed on pressures caused by expanding water and ordered water, as well as by crystallization and hydration pressures of entrapped salts. The process progresses in very small openings remote from human observation; thus controlled experiments would not separate osmosis from other processes. The diagrams offer a theoretical evaluation of the highest possible pressure developed by osmosis. The following environmental and geological implications appear to be involved: rock temperatures

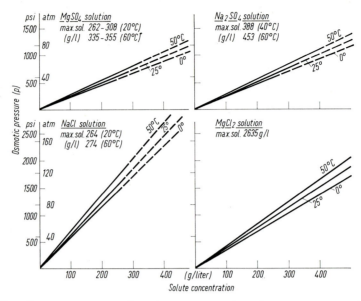

Fig. 92. Osmotic pressure versus the solute concentration of some common sulfates and chlorides

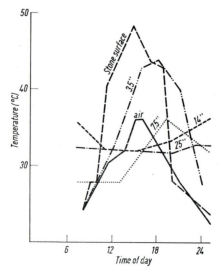

Fig. 93. Diurnal temperature distribution in granitic boulder of near Barstow, California, Mojave Desert. Note that surface temperature of rock well exceeds the air temperature. Data plotted from ROTH (1965)

in cities, diurnal temperature changes, mole concentration of pore waters in stone, and the question of membranes.

Rock temperatures: Rock temperatures in deserts and on stone buildings in city climates may readily reach 50° C or more close to the stone surface. ROTH's (1965) temperature profile across a desert boulder of granitic rock is plotted in Fig. 93: temperatures inside a thick masonry wall may well resemble the temperature lag in a rock boulder or dimension stone. Temperature maxima are plotted for different depths beneath the stone surface. Fig. 94 plots the average temperature gradient against the maximum and the minimum temperature of a hot summer day in the Mojave Desert, California, on the 35th parallel. The temperature at the rock surface is about 33% higher than the maximum air temperature. A black gabbro, for instance, is believed to reach a temperature of about

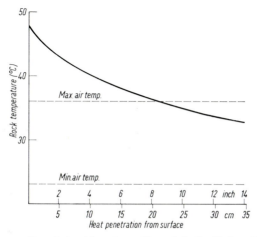

Fig. 94. Temperature gradient in boulder of granitic rock in the Mojave Desert near Barstow, California. Data are for August 22, 1962, recorded by ROTH (1965) and plotted by this author

55° C. Similar temperature conditions should be expected on vertical stone walls, especially early in fall when the sun is able to hit vertical walls at a less acute angle.

Salt concentrations in closed pore systems have given rise to much speculation, because the molar concentration of salts in closed pore systems is very little known. In general, entrapped salts are not expected to form concentrations in excess of the maximum solubility of the dissolved ions. KAISER's (1929) cross section through a corner stone from the Regensburg Cathedral suggests that the content of soluble salts in a sandstone does not exceed 5%.

Resistance to osmotic pressures should not affect the tensile strength of sound stone which qualifies as construction material. In weathered rock the tensile strength may drop near 100 atmospheres in which HEIDECKER (1968) could observe spalling. Surface spalling is an important component in salt weathering, generally summarized as "salt fretting".

Osmotic conditions are believed to exist within partly weathered igneous rocks and possibly in isolated vugs of carbonate rocks.

1. In partly weathered igneous rocks, gels of silica and alumina are believed to form semi-permeable membranes in the sense of Liesegang. Scaling and spalling associated with spheroidal weathering in igneous rocks may be explained by osmosis.

2. Isolated vugs in carbonate rocks, mostly dolomites, can concentrate carbonates of Ca and Mg in solution, also sulfates of Mg. This author does not believe that enough salts can be concentrated within the vug to develop destructive osmotic pressures.

3. Ordered water and surface-adsorbed exchangeable ions can create osmotic conditions in argillaceous carbonates and shales; these appear to be related to soil osmosis which is responsible for much of the swelling of clays. Both non-swelling and swelling clay crystals can hold water at their surface so tightly that it becomes ordered in micropores less than 10 millimicrons in diameter. The great rigidity of such ordered water can well function as an ionic screen or membrane because this water is much denser and about 20 times more viscous than ordinary water and it does not freeze above $-40°$ C. Ions are readily adsorbed to clay crystals; these

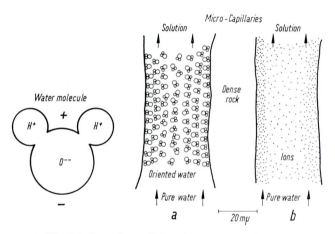

Fig. 95. Osmotic conditions in narrow capillaries:

a Ordered water as a semi-permeable membrane
b Ions held tightly on mineral surfaces (clay) as membrane

can be exchanged by the mechanism of base exchange of one ion against another. The capacity of ion exchange is expressed in milliequivalents (me) per 100 g. of clay substance. The ion density decreases exponentially with increasing distance from the pore wall. Osmosis is well known in soil systems whereby the osmotic pressure is inversely proportional to the initial water content. Suction occurs towards areas of high ionic concentration where dilution ensues and subsequent swelling of the soil-water system. Clay may occur in dolomites as interconnected systems of bands through rejection by the recrystallizing dolomite grains during the process of lithification (DUNN and HUDEC, 1966). Such osmotic conditions are illustrated in Fig. 95. Swelling pressures in carbonate rocks with a clay content should not exceed 50 atm according to KOMORNIK and DAVID (1969). Soils over carbonate rocks reflect the entire insoluble residue of these rocks. Montmorillonite, the only swelling common clay mineral, is relatively sparse in residual soils.

In the field, the source for osmotic pressures in stone and concrete is difficult to assess, as the various factors which lead to osmosis overlap. On the other hand, osmosis overlaps with other disruptive forces such as frost action and expanding water.

6.8. Efflorescence on Stone and Brick Surfaces

Blotches, patches and margins of white salts on masonry and brick form unpleasant-looking coatings. Soluble salts crystallize at the surface at the open ends of capillary systems where outward-moving moisture is the vehicle of transport. The mechanism of efflorescence is sketched in Figs. 96 and 97. The fluid motion during evaporation and subsequent crystallization is marked by arrows. The fluid

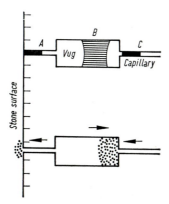

Fig. 96. Mechanism of efflorescence on stone surface from capillary and vug combination. After LAURIE and MILNE (1926)

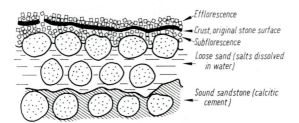

Fig. 97. Crystallization of salts beneath and above the stone surface as efflorescence and subflorescence. After SCHMIDT (1969)

migrates from the narrow capillary towards a vug and from there to the surface. Salts on the surface generally do not harm the masonry but rather inform us of salt migration and upper fringes of moisture travel. The mechanism of efflorescence is discussed by many authors, but was summarized best by KIESLINGER

(1957, 1963). Brick walls show this phenomenon well by the color contrast from the white salts to the red brick background. Sulfates from curing and firing of the clay supplies more sources for efflorescence in bricks than in stone masonry.

Fig. 98. Discoloring of fine-grained granite by efflorescence. Moisture travelled along mortar joints. Source of salts may be the mortar. Inside of small mausoleum, South Bend, Indiana

Fig. 99. Concentration of moisture near the joints of the granite panels, a possible warning of future efflorescence, subflorescence or other signs of damage. Infiltration of the dark wet spots (arrows) with salts should be expected. Mural Mosaic of the Notre Dame Memorial Library

LAMAR and SCHRODE (1953) found the sulfates of Ca and Mg to be most common among water-soluble salts in stone. Although rarely in excess of 0.7% of the total rock substance these salts lead to efflorescence and subflorescence in quarries and natural rock outcrops. In stone buildings mortar joints and ground moisture add soluble salts to residual solubles from the quarries. Fig. 98 pictures the interior of a small mausoleum; fine-grained New England granite is marred by efflorescent margins which run parallel to the mortar joints. Wet margins along granite slabs may be seen on the lower part of the stone mural on the south side of the Notre Dame Memorial Library (Fig. 99). No discoloring other than by moisture is visible yet. The mortar joint appears to be the present travel route for most of the moisture which entered by driving rains.

6.9. Subflorescence

Subflorescence is closely related to efflorescence; the salts move towards the stone surface as in efflorescence without reaching the surface itself. Instead they crystallize beneath a crust of weathered rock substance, dust and soot forming a thin, indurated skin. The skin loses its support and starts peeling; the surface is ready for another skin which again will peel as a result of surface induration by salt concentration at the surface and subsequent salt action beneath the skin. Fig. 97 sketches a cross section through a fine-grained sandstone which has developed both near-surface solution of the calcareous cement and surface concentration leading to efflorescence and subflorescence. Deserts and urban humid climates are favorable environment for subflorescence.

6.10. Crystallization Pressure of Salts in Stone

The crystallization process of salts from solutions is known to produce pressures by nonaccommodative crystal growth in small pores of rock which leads to disruption. EVANS (1969/70) reviewed the literature on salt crystallization. The most important reference cited is CORRENS (1949); he summarizes the state of knowledge and presents a workable equation based on the Riecke principle. A crystal under linear pressure has greater solubility than an unstressed crystal.

$$P = \frac{R \cdot T}{V_s} \cdot ln C/C_s$$

P is the pressure by crystal growth;
R is the gas constant of the ideal gas law, 0.082 liter-atmospheres per mole degree;
   T is the temperature in degrees Kelvin;
   V_s is the molecular volume of the solid salt;
   C is the actual concentration of the solute during crystallization;
   C_s is the concentration of the solute at saturation.

A state of supersaturation is essential for the crystallization. Fig. 100 plots the temperature T of the solution against the concentration ratio of the solution C/C_s on an indefinite graph (MULLIN, 1961). Crystallization can only occur between the states of saturation and supersaturation. No crystallization takes place in unsaturated solutions. Three theoretical situations are plotted in the graph:

a) A to B: At constant concentration the temperature drops from A towards B during a daily temperature cycle and the zone of saturation is entered. Supersaturation is thus reached and crystallization can occur in this metastable state. With higher concentration of the solution, crystallization is more rapid and the crystals grow larger than with lower concentrations.

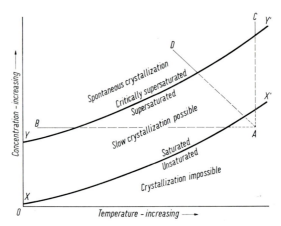

Fig. 100. Indefinite temperature-solute concentration plot. Supersaturation can be obtained by cooling only (A to A′), concentration increase (A to C), and a combination of both (A to D). Modified from MULLIN (1951)

b) A to C: The concentration rapidly increases with constant temperature by evaporation from the ends of capillaries, but can also revert to a lower concentration by dilution through rainwater.

Table 25. *Crystallization Pressures for Some Salts*

Salt	Chemical Formula	Density (gms/cm.^{3})	Molecular Weight (gms/mole)
Anhydrite	$CaSO_4$	2.96	136
Bischofite	$MgCl_2 \cdot 6\ H_2O$	1.57	203
Dodekahydrate	$MgSO_4 \cdot 12\ H_2O$	1.45	336
Epsomite	$MgSO_4 \cdot 7\ H_2O$	1.68	246
Gypsum	$CaSO_4 \cdot H_2O$	2.32	127
Halite	$NaCl$	2.17	59
Heptahydrite	$Na_2CO_3 \cdot 7\ H_2O$	1.51	232
Hexahydrite	$MgSO_4 \cdot 6\ H_2O$	1.75	228
Kieserite	$MgSO_4 \cdot H_2O$	2.45	138
Mirabilite	$Na_2SO_4 \cdot 10\ H_2O$	1.46	322
Natron	$Na_2CO_3 \cdot 10\ H_2O$	1.44	286
Tachhydrite	$2\ MgCl_2 \cdot CaCl_2 \cdot 12\ H_2O$	1.66	514
Thenardite	Na_2SO_4	2.68	142
Thermonatrite	$Na_2CO_3 \cdot H_2O$	2.25	124

c) A to D: Change of both the temperature and the concentration leads to supersaturation where crystallization occurs.

Crystallization pressures were calculated and plotted by WINKLER and SINGER (in press) as three-dimensional diagrams utilizing CORRENS's equation as the most practical thermodynamic approach. It correlates well with laboratory tests. Complete curves were drawn for gypsum and halite, also for sodium sulfate which is claimed most efficient in rock disintegration by KWAAD (1970); see Figs. 101,

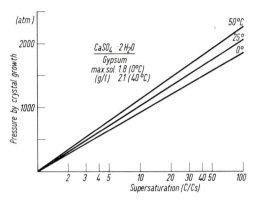

Fig. 101. Plot of crystallization pressures of gypsum, for various temperatures and degrees of supersaturation. From WINKLER and SINGER (1972)

102; the degree of supersaturation is plotted on the abscissa on a log scale, the pressure generated by crystal growth on the ordinate on a linear scale. The pressure-concentration relationship is a straight line with the natural log of C/C_s for a given temperature: lines of equal temperature are drawn for 0°, 25°, and 50° C.

Table 25 *(Continued)*

| Molar Volume (cm.³/mole) | Crystallization Pressure (atm) | | | | | |
| | $C/C_s = 2$ | | $C/C_s = 10$ | | $C/C_s = 50$ | |
	0° C	50° C	0° C	50° C	0° C	50° C
46	335	398	1120	1325	1900	2262 ·
129	119	142	397	470	675	803
232	67	80	222	264	378	450
147	105	125	350	415	595	708
55	282	334	938	1110	1595	1900
28	554	654	1845	2190	3135	3737
154	100	119	334	365	568	677
130	118	141	395	469	671	800
57	272	324	910	1079	1543	1840
220	72	83	234	277	397	473
199	78	92	259	308	440	524
310	50	59	166	198	282	336
53	292	345	970	1150	1650	1965
55	280	333	935	1109	1590	1891

The maximum saturation on the diagrams is $C/C_s = 100$, which is the critical supersaturation for many salts; beyond this point no salt is left in solution. The curves are given for ideal conditions of crystallization which are almost never found in nature. Table 25 summarizes calculations for salts which are common ingredients

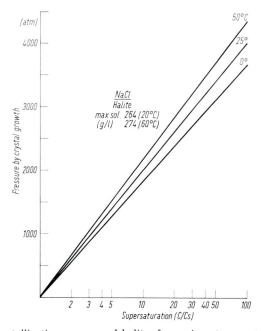

Fig. 102. Plot of crystallization pressure of halite, for various temperatures and degrees of supersaturation. From WINKLER and SINGER (1972)

in pores of rock, stone and concrete. The pressures were calculated from the density, the molecular weight and the molar volume for supersaturations C/C_s of 2, 10, and 50. At C/C_s equals 10, halite can exert a maximum pressure of 2190 atm; this pressure is sufficient to disrupt the strongest rock.

6.11. Hydration Pressure

Some common salts readily hydrate and dehydrate in response to changes in the temperature and relative humidity of the atmosphere; adjustment to the new environment takes place as the salt changes to a more stable hydrate. The absorption of water increases the volume of the salt and thus develops pressure against the pore walls. The geomorphologist MORTENSEN (1933) first recognized the importance of salt hydration as an important factor in desert weathering and attempted the calculations of hydration pressures. Monuments in cities of moderate semi-humid climate may be readily saturated with salts introduced into stone from polluted air, mortar and groundwater. WINKLER and WILHELM (1970) calculated the hydration pressures of some important common salts at different temperatures and relative humidities. The following equation was developed:

$$P = \frac{(n \cdot R \cdot T)}{V_h - V_a} \times 2.3 \log \frac{P_w}{P_w{}'}$$

where P is the hydration pressure in atmospheres,

n is the number of moles of water gained during hydration,

R is the gas constant of the ideal gas law, in 82.07 milliliter-atmosphere per mole-degree,

T is the absolute temperature in degree Kelvin,

V_h is the volume of the hydrate, in cubic centimeter per gram-mole of hydrate salt, it equals the mole-weight of hydrate over the density,

V_a is the volume of the original salt before hydration, in cubic centimeter per gram-mole, it equals the mole weight of the original salt over the density,

P_w is the vapor pressure of water, in mm. mercury at given temperature, and

$P_w{}'$ is the vapor pressure of hydrated salt, in mm. mercury at given temperature.

Curves are drawn as three-dimensional contours within the fields of stability for each salt. The abscissa gives the relative humidity of the atmosphere in contact with the salt, the ordinate the pressure P in atmospheres, called the hydration pressure; contours are drawn from 0° to 60° C, at 10° intervals; these lines permit the exact theoretical determination of the maximum pressure exerted during the hydration process at a given relative humidity and temperature of the salt. At 0° C and 90% relative humidity the maximum pressure of gypsum during hydration is about 2000 atm, at 30° C only about 1480 atm, and so forth.

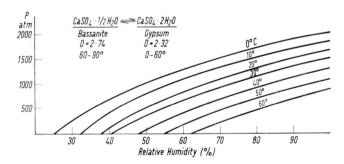

Fig. 103. Plot of hydration pressure from the hemihydrate bassanite to the dihydrate gypsum, for various temperatures and relative humidities. From WINKLER and WILHELM (1970)

The pressure-temperature-humidity curves are theoretical and present maxima under idealized conditions of closed-pore systems, unlimited moisture access towards the hydrating crystals in the pores, etc. Indicated temperature stability ranges of the hydrated minerals are theoretical, as metastable conditions may exist for a long time at much lower temperatures than indicated. Some of the calculations are extended to temperatures where metastable conditions may well exist. Calculations were performed with the sulfates of Ca, Na, Mg, and the carbonate of Na. Fig. 103 presents the complete pressure-relative humidity-temperature diagram for the hydration of plaster of Paris, the hemihydrate of calcium sulfate, to the dihydrate, gypsum. Table 26 summarizes the pressures for several common salts. In general, low temperatures and high relative humidities produce the highest pressures, high temperatures and low relative humidities low pressures.

Table 26. *Hydration Pressures of Some Common Salts, Simplified*

relat.
Humidity Temperature (° C)
(%)   0°   20°   40°   60°

$CaSO_4 \cdot \frac{1}{2} H_2O$ to $CaSO_4 \cdot 2 H_2O$ (plaster of Paris to gypsum)

(%)	0°	20°	40°	60°
100	2190	1755	1350	926
90	2000	1571	1158	724
80	1820	1372	941	511
70	1600	1145	702	254
60	1375	884	422	0
50	1072	575	88	

$MgSO_4 \cdot H_2O$ to $MgSO_4 \cdot 6 H_2O$
Kieserite to Hexahydrite
65.3° C

$MGSO_4 \cdot 6 H_2O$ to $MgSO_4 \cdot 7 H_2O$
Hexahydrite to Epsomite

(%)	65.3° C		10°	20°	30°	40°
100	418		146	117	92	96
90	226		132	103	77	69
80	13		115	87	59	39
70			97	68	40	5
60			76	45	17	
50			50	19	0	
40			20	0		

$Na_2CO_3 \cdot H_2O$ to $Na_2CO_3 \cdot 7 H_2O$
Thermonatrite to Heptahydrite

$Na_2CO_3 \cdot 7 H_2O$ to $Na_2CO_3, 10 H_2O$
Heptahydrite to Natron

(%)	0°	10°	20°	30°	0°	10°	20°	30°
100	938	770	611	430	816	669	522	355
90	799	620	457	276	666	504	350	185
80	637	455	284	94	490	320	160	0
70	448	264	88		282	112	0	
60	243	46			60			

Data compiled by WINKLER and WILHELM (1970).

The hydration of the hemihydrate plaster of Paris has been observed to be a rapid process, whereas a very slow process from anhydrite to the hemihydrate. ENGELHARDT's (1945) experiments with anhydrite powder are in contrast with the previous statement and general belief: anhydrite powder readily recrystallizes to gypsum in the laboratory within a week. The hydration pressure of anhydrite to gypsum without passing the intermediate stage of plaster of Paris, may produce a maximum pressure of 2800 atm at 0° C at 100% relative humidity. The hydration of the various sulfates of magnesium — kieserite to hexahydrite, to epsomite and to the dodecahydrate — is limited in nature and exerts only low pressures. Hexahydrite was reported by FOSTER and HOOVER (1963) as a common efflorescent salt in dolomite stone quarries of northern Ohio, where the metastable hexahydrite had crystallized from solution at the surface at temperatures well below the minimum of its stability. MORTENSEN (1933) has observed the hydration of the metastable kieserite to epsomite to happen in a single day.

The three hydrates of sodium carbonate are common in nature. Sufficient stress may develop for all hydrates to destroy stone. The hydration of the sodium sulfates thenardite to mirabilite is more rapid than hydration of other salts; their hydration and dehydration may repeat several times in a single day: even low hydration pressures may become effective in such rapid change. The dehydration of mirabilite to thenardite does not take longer than 20 minutes at 39° C.

A full evaluation of the behaviour of salts in very narrow capillaries is very difficult.

6.12. Differential Thermal Expansion of Salts

Disruption of stone may also take place by reason of the considerable contrasts in thermal expansion of entrapped salts in the pores of stone. COOKE and SMALLEY (1968) assign great importance to the high degree of expansion of halite at temperatures below 100° C as compared with the surrounding rock substance.

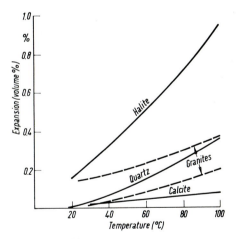

Fig. 104. Differential expansion of halite as compared with quartz, calcite, and granite. Adapted from COOKE and SMALLEY (1968)

Fig. 104 plots thermal expansions of calcite, quartz, granite, and halite. Upon temperature increase from near freezing to 60° C halite expands 0.5% whereas granite less than 0.2%. Small as this difference may appear, it is believed that disruptive expansion of salt may participate in physical stone decay.

6.13. Salt Fretting

Salt fretting or salt erosion embraces one or several salt actions without differentiating among them. A few examples are cited in the following:

1. Street salts: Winter street salting results in an excess supply of halite or $CaCl_2$ on streets and sidewalks to melt the ice and improve the traffic safety. Infiltration of such salts into stone has caused extensive damage. Steps of coarse-grained Georgia marble at the Shedd Aquarium in Chicago have started cracking and bloating; numerous radial cracks along week bands made partial replacement of the stone necessary. The stone was set 40 years ago. Fig. 105 gives an oblique view of a step.

Beneath pillars of Indiana Limestone fine-grained Massachussetts grey granite is flaking along the outer edges in front of the Mayflower Hotel in Washington,

Fig. 105. Cracking and swelling of coarse-grained Georgia marble on much travelled stairway. The intersection of the banding with the surface accentuates less resistant bands. Moisture and salt action (halite) appear to be most responsible for damage. Entry to Shedd Aquarium, Chicago, Illinois. From WINKLER and SINGER (1972)

D.C. Cause of the damage should be sought in the effect of street salting, as the damage to the granite is only visible on both sides of the entrance to the bar of the hotel (Fig. 106).

Honeycomb weathering in quartzitic sandstone developed above the main entrance of Sacred Heart Church on the Notre Dame campus. A combination of high-rising street salts, drenching with driving rains and drying rapidly by direct exposure to the sun, may be responsible for the damage (Fig. 107).

2. Salts in ground moisture have, no doubt, concentrated in the masonry of many older buildings which lack proper insulation to ground moisture. Figs. 108 and 109 picture a sequence of two photos of the development of deep honeycombs

Fig. 106. Etched Indiana limestone on white granite, separated by mortar joint. Granite spalls at corners (arrows). Origin of damage to granite should be sought in salt action from street salting or pollution-converted calcium sulfate from the Indiana limestone which is strongly etched. Mayflower Hotel, Washington, D.C. Pillars do not show damage to granite, except at the bar entrance, where street salting was apparently intensified. From WINKLER and SINGER (1972)

Fig. 107. Honeycomb weathering in quartzitic sandstone. Winter salting of sidewalk provides salt, transported by action of driving rain and sun on south side of Sacred Heart Church, Notre Dame, Indiana. The church is 100 years old. Photo 1964

Fig. 108. Ground moisture and efflorescence destroyed soft fine-grained shell limestone with quartz grains (Miocene Molassae) from quarries at Angles (Gard). Stone is strongly honeycombed near ground surface. Photo was taken just before remodelling in 1938. 17th century chapel of the Penitents in Avignon, France. Photo from Dr. Jaton, Ministere des Affaires Culturelles, France (Cl. No. 92.475)

Fig. 109. Damage by ground moisture and strong efflorescence has developed again after replacement of damaged stone shown in Fig. 108. Photo was taken 10 years after stone replacement. South Facade of Penitents Chapel, Avignon, France. Photo from Dr. Jaton, Ministere des Affaires Culturelles, France (Cl. No. 92.475)

near the base of the Penitent Chapel in Avignon, France. Deep gouging is evident to a height about 6 feet above the ground surface; after replacement of most damaged stones in 1938, damage was evident again 10 years later. Salt action has nearly destroyed the famous Romanesque portal of the monastery Santa Maria at Ripoll, in the Pyrenees of Spain. Sulfates and chlorides of Na, Ca and K were the most important water-soluble ions trapped in the stone, a porous limestone. Cleaning of the monument, chemical treatment and proper ground drainage are believed to have stopped further decay (CABRERA, 1966).

In a series of systematic laboratory experiments KWAAD (1970) found very different reaction of salts on granite. Sodium sulfate was the most effective disintegrating factor on granite, followed by sodium carbonate and magnesium sulfate; gypsum was the least effective. The damage in granite from all of these was purely mechanical.

The ASTM sulfate soundness test (see Appendix B) shows more attack by $MgSO_4$ than by Na_2SO_4 although both salts exert about the same crystallization pressure for the single hydrate. The greater attack appears to be due to the much faster hydration rate of magnesium sulfate than of sodium sulfate.

6.14. Rust Stains and Desert Varnish

Some white marbles, grey granites and other light-colored rocks occasionally develop objectionable rusty stains at the surface. Iron, often with some manganese, can be mobilized during the weathering process from decaying pyrite in the stone, from hooks and iron braces, and also from the ground by means of capillarity. Surface stains of iron bear great similarity to the well-studied desert varnish (ENGEL and SHARP, 1958). The mechanism of iron mobilization is discussed in more detail in the chapter on Natural Rust. The formation of natural desert varnish is a slow process; there is evidence that this process may take 25 years in nature, possibly less in urban areas. KIESLINGER (1932) reports discoloring of the formerly white Pentelic marble of the Acropolis of Athens: the solution-oxidation process of tiny white grains of siderite (iron carbonate) present in the marble may possibly be responsible for the color change.

6.15. Surface Induration, Frames, and Honeycombs

The movement of moisture towards the stone surface and towards outer edges results in the formation of hard crusts parallel to the worked surface. Such crusts may be quite protective if composed of secondary calcite, calcitic sinter, or of calcium sulfite. Superficial induration remains temporary if the disintegrating stone substance is the sole supplier of the ingredients for the outer crust; the interior structure of the stone block may weaken to such an extent that sandstones may be entirely deprived of their grain cement leading to collapse of the stone. KIESLINGER's (1959) classic paper on frame weathering (Rahmenverwitterung) discusses many phases and stages of stone destruction by longer exposure to moisture from within the center of stone slabs, together with faster drying along the edges towards the mortar joints. Surface induration of stone surfaces is well known to occur by salt transport from the inside towards the outside (Fig. 110). Volcanic trachyte showed signs of interior decay by salt action on the Cathedral of Colon after only 30 years of exposure (KNETSCH, 1952). Sandstones with calcitic cement and impure carbonate rocks are endangered most by frame weathering.

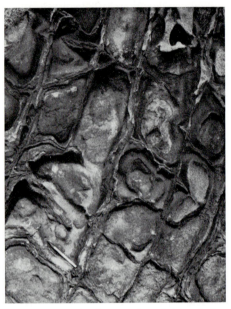

Fig. 110. Spheroidal weathering: casehardening along joints darkened by limonite concentration. Slab of quartzitic sandstone from Mt. Sinai, Israel, now set up for display at the Morris Greenhouse, South Bend, Indiana. This spectacular example for desert weathering was not modified by the semihumid climate of South Bend

Table 27. *Physical and Chemical Profile across a Calcareous Sandstone*

	Calcitic grain cement %	Sulfate %	Porosity %
Sound sandstone	100	0	100
Transition from sound to weathered	85	5	120
Loose sand, cement dissolved	0	0	150
Subfluorescent crust	300—500	10—100	50
Outer hard crust	300—0	100—0	10

Data from: SCHMIDT (1969).

SCHMIDT (1969) generalizes surface induration in a sandstone by the proportion of the calcitic grain cement, the stone porosity and the sulfate content; Fig. 97 pictures a cross section; the data are presented in Table 27. The original porosity is assigned a value of 100.

Fig. 111. Honeycomb weathering in red Fountain arkose sandstone, in semi-arid climate.
Little Fountain Canyon, Colorado Springs, Colorado

Fig. 112. Cleopatra's Needle, New York City. The east face appears undamaged while the
south face shows transition from poorly preserved towards destruction. From WINKLER (1965)

KNETSCH (1960), KNETSCH and REFAI (1965), as well as VOÛTE (1963) provide ample information on frame weathering from the Egyptian desert. The formation of honeycombs in sandstones and limestones are characteristic arid forms of weathering which are not infrequently found on buildings in humid and semi-humid urban climates. Rapid migration of solutions combined with micro-climatic differences along channels may provide weathering along channels leading to a honeycomb-like structure. Figs. 111 and 107 compare the honeycombed red coarse-grained Fountain sandstone near Colorado Springs, Colorado, with a grey, quartzitic sandstone above the entrance of Sacred Heart Church on the campus of the University of Notre Dame; there the stone had been exposed to both driving rain from the southwest and intense exposure to sunlight for only 100 years.

6.16. Urban and Desert Climate, a Comparison

Urban climates have been regarded as milder than rural environs; actually urban climates are the more severe on stone and concrete. The reflection of solar radiation against rows of houses in narrow streets causes snow to disappear from urban areas first, winds are less strong, fog more frequent, etc. LOWRIE (1967) summarizes the information which was obtained for many major American cities in semi-humid and humid climates of the middle latitudes and compares them with the rural areas. LOWRIE's chart of Fig. 113 presents information which is

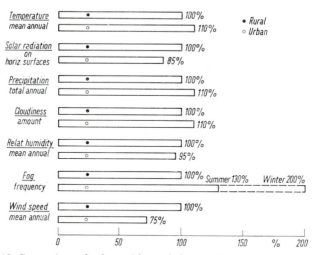

Fig. 113. Comparison of urban with rural climate. Data from LOWRY (1967)

pertinent for the understanding of urban stone decay. The data presented in the chart have to be accepted with some caution, because the distribution of housing, high-rise buildings, park areas, etc., plays an important role in the final evaluation of the site under investigation. Temperatures, solar radiation, relative humidity, amount of fog, cloudiness and wind speeds influence the urban climate:

a) Temperature is 10% higher by multiple light reflection and numerous heat sources despite actual solar radiation of less than 15%.

b) Solar radiation for horizontal surfaces is reduced in cities about 15% because more cloudiness, smoke, smog and industrial dust obscure the sun more frequently. Reduced exposure of exterior stone surfaces to sunlight prevents satisfactory drying of surfaces.

c), d) Both total cloudiness and rate of precipitation are about 10% higher in cities due to a great abundance of condensation nuclei from dust and smoke. Less sunshine results in slower drying of wet stone surfaces.

e) The annual mean relative humidity is about 5% lower in cities than in rural areas where lower average temperatures, more transpiration over woodland, open bodies of water and fields provides more moisture. Relative humidity is an important factor in the hygroscopic sorption of water. The alternate heating and cooling of stone surfaces still provides sufficient condensation to inflict damage.

Table 28. *Comparison of Weathering in Deserts with Weathering in Cities*

Kind of Weathering	Desert	Cities
Damaging Agents:		
temperature contrasts	high	high (on walls)
moisture from ground	high	high
moisture from fog	fog in winter	frequent
moisture from condensation	present	present
salts, origin of salts	desert floor, residual	groundwater, stone weathering, polluted air
Damage to Stone:		
abrasion by wind action	strong	some, near street level
frost action	occasional	high
flaking by heat and moisture	very strong	moderate
efflorescence	strong	moderate but common
subflorescence	strong	moderate to strong
frame weathering, case hardening	common as hollow pebbles	common
desert varnish and stains	hard brown crusts	light-brown stains
solution	very slow	very rapid in polluted atm.

f) Fog develops as the atmosphere reaches the dew point at 100% relative humidity. Fog is the most important type of precipitation in stone decay as the tiny fog droplets can remain suspended in the atmosphere for many hours where they are able to scavenge the atmosphere of suspended pollutants and slowly react with stone. The presence of fog appears to be more destructive to stone and concrete than many deleterious ingredients.

g) The annual mean wind speed in urban areas is reduced 25% through numerous wind breaks in a city.

Urban areas compare unfavorably with their rural environs as more moisture is provided with less possibility to dry wet stone surfaces. Higher stone temperatures prevail in cities. No data are available which permit a quantitative comparison between both environments nor are the variables known well enough to calculate a theoretical model for the destruction of stone in urban areas.

Stone decay in humid urban climate and deserts: Stone decay in urban humid areas of the middle latitudes has great similarity with decay in desert areas. Urban climate permits good correlation with deserts as salt action and temperature contrasts play an equally great importance on stone. Table 28 compares the two types of weathering.

Summary

Moisture in stone is of the greatest importance, both as a vehicle of transport for salts and as a disruptive agent. Capillary travel within stone is extensive towards areas where it is often not expected. Upward movement of moisture may reach 20 m. above the ground, horizontal travel twice as much. Expanding moisture in pores may develop as much as 7500 psi when heated from near freezing to 60° C. Observed spalling and surface flaking should be merely ascribed to the expansive action of pure water. The disruptive action of ordered water is little known nor understood.

Salts which may be derived from ground moisture, stone weathering, or polluted air travel readily in stone both by diffusion and by capillarity. Surface efflorescence, near-surface subflorescence, salt crystallization and salt hydration in stone pores can produce considerable pressures, in excess of 30,000 psi for halite and gypsum. Differential expansion of entrapped salts during stone heating is probably less important. Stone disruption by salt action is destructive because low tensile strengths of rock amount to only one-fifth to one-tenth the compressive strength. Osmotic pressures may affect partly weathered silicate rocks and soft shales.

Weathering in deserts is strikingly similar to weathering conditions in urban areas. The process of damage due to moisture, with or without salts, is extremely complex and thus very difficult to analyze quantitatively.

References

1. BRUECKNER, W. D., 1966: Salt weathering and inselbergs. Nature, **210**, 832.
2. BURGESS, S. G., and R. J. SCHAFFER, 1952: Cleopatra's Needle. Chemistry and Industry, **1952**, 1026—1029.
3. CABRERA-GARRIDO, J. M., 1968: Le Portail du Monastère de Santa-Maria de Ripoll. Étude Scientifique pour la Conservation. Colloques sur L'Altération des Pierres, Vol. 1, Bruxelles 1966—1967, pp. 127—165. Conseil International des Monuments et des Sites, ICOMOS, Palais de Chaillot, Paris.
4. CARL, J. D., and G. C. AMSTUTZ, 1958: Three-dimensional Liesegang rings by diffusion in a colloidal matrix. Geol. Soc. America, Bull., **69** (**4**), 1467—1468.
5. CHOQUETTE, P. W., and L. C. PRAY, 1970: Nomenclature and classification of porosity in sedimentary carbonates. Am. Assoc. Petroleum Geologists, Bull., **54** (**2**), 207—250.
6. COOKE, C. W., and I. J. SMALLEY, 1968: Salt weathering in deserts. Nature, **220**, 1226 to 1227.
7. CORRENS, C. W., 1949: Growth and dissolution of crystals under linear pressure. Discussions of the Faraday Society, **5**, 267—271.

8. DUNN, J. R., and P. P. HUDEC, 1966: Water, clay, and rock soundness. Ohio Journal of Science, **66** (**2**), 153—168.

9. ENGEL, C. G., and R. P. SHARP, 1958: Chemical data on desert varnish. Geolog. Society America Bull., **69** (**2**), 487—518.

10. ENGELHARDT, W. v., 1945: Zur Bildung von Gips von Anhydrit. Chemie der Erde, **XV**, 424—428.

11. EVANS, I. S., 1969/70: Salt crystallization and rock weathering: a review. Revue de Geomorphologie Dynamique, **XIX** (**4**), 153—177.

12. FOSTER, R. F., and K. V. HOOVER, 1963: Hexahydrite ($MgSO_4 \cdot 6\,H_2O$) as an efflorescence of some Ohio dolomites. Ohio Jour. of Science, **63** (**4**), 152—158.

13. GARRELS, R. M., R. M. DREYER, and A. L. HOWLAND, 1949: Diffusion of ions through intergranular spaces in water-saturated rocks. Geolog. Soc. America Bull., **60** (**12**), 1809—1828.

14. GRIGGS, D. T., 1936: The factor of fatigue in rock exfoliation. Jour. of Geology, **44** (**7**), 783—796.

15. HEIDECKER, E., 1968: Rock pressures induced by weathering and physiochemical processes. Proc. Australian Inst. Mining and Metallurgy, **226** (**1**), 43—45.

16. KAISER, E., 1929: Über eine Grundfrage der natürlichen Verwitterung und der chemischen Verwitterung der Bausteine in Vergleich mit der in der freien Natur. Chemie der Erde, **IV**, 291—342.

17. KIESLINGER, A., 1932: Zerstörungen an Steinbauten, ihre Ursachen und ihre Abwehr. Leipzig-Wien: Deuticke.

18. KIESLINGER, A., 1957: Feuchtigkeitsschäden an Bauwerken. Zement und Beton, **9**, 1—7.

19. KIESLINGER, A., 1959: Rahmenverwitterung. Geologie und Bauwesen, **24** (**3—4**), 171—186.

20. KIESLINGER, A., 1963: Verwitterungseinflüsse an Ziegelmauerwerk. Die Wienerberger, **3**, 3—9.

21. KNETSCH, G., 1952: Geologie am Kölner Dom. Geol. Rundschau, **40** (**1**), 57—73.

22. KNETSCH, G., 1960: Über aride Verwitterung unter besonderer Berücksichtigung natürlicher und künstlicher Wände in Ägypten. Zeitschr. f. Geomorphologie, Supplementum I, Morphologie des Versants, pp. 190—205.

23. KNETSCH, G., und E. REFAI, 1955: Über Wüstenverwitterung, Wüstenfeinrelief und Denkmalzerfall in Ägypten. Neues Jb. Geologie u. Paläontologie, Abh., **101**, 227—256.

24. KOMORNIK, A., and D. DAVID, 1969: Prediction of swelling pressure of clays. J. of Soil Mechanics and Foundation Division. Proc. Soc. Civil Engineers, Jan. 1969, pp. 209—225.

25. KWAAD, F. J. P. M., 1970: Experiments on the granular disintegration of granite by salt action. From Field to Laboratory, Publ. # 16, Univ. of Amsterdam, Holland, Fysisch Geografisch en Bodemkundig Laboratorium, pp. 67—80.

26. LAMAR, J. E., and R. S. SCHRODE, 1953: Water soluble salts in limestones and dolomites. Economic Geology, **48**, 97—112.

27. LAURY, A. P., and J. MILNE, 1926: The evaporation of salt solution from surfaces of stone, brick and mortar. Proc. Royal Soc. of Edinburgh, **47**, 1926—1928, 52—68.

28. LIESEGANG, R. E., 1945: Geologische Bänderung durch Diffusion und Kapillarität. Chemie der Erde, **XV**, 420—423.

29. LOWRIE, W. P., 1967: The climate of cities. Scientific American, **217** (**2**), 15—23.

30. MAHAN, B. H., 1964: Elementary chemical thermodynamics. Amsterdam: W. A. Benjamin, Inc., 155 p. (paper back).

31. MAMILLAN, M., 1968: L'altération et la préservation de pierres dans les monuments historiques. Etude de L'Altération des Pierres, Vol. 1, Colloques tenus à Bruxelles le Férr., 1966—1967, pp. 65—98. Conseil International des Monuments et des Sites, ICOMOS, Palais de Chaillot, Paris.

32. MORTENSEN, H., 1933: Die Salzsprengung und ihre Bedeutung für die regional klimatische Gliederung der Wüsten. Petermann's Geographische Mitteilungen, pp. 130—135.

33. MULLIN, J. W., 1961: Crystallization. London: Butterworths, 268 p.

34. NUR, A., and G. SIMMONS, 1970: The origin of small cracks in igneous rocks. Internat. Journal Rock Mechanics and Mining Sciences, **7** (**3**), 307—314.

35. ROTH, E. S., 1965: Temperature and water content as factors in desert weathering. Journal Geology, **73** (**3**), 454—468.

36. RUIZ, C. L., 1962: Osmotic interpretation of the swelling of expansive soils. Highway Research Board, Bulletin 313, pp. 47—77.

37. SCHMIDT-THOMSEN, K., 1969: Zum Problem der Steinzerstörung und Konservierung. Deutsche Kunst und Denkmalpflege, pp. 11—23.

38. VOS, B. H., and E. TAMMES, 1968: Flow of water in the liquid phase. Rept. No. B 1-68-38, Inst. TNO for Building Materials and Building Structures, Delft, Holland, 45 p.

39. VOS, B. H., and E. TAMMES, 1969: Moisture and moisture transfer in porous materials. Rept. Nr. B 1-69-96, Inst. TNO for Building Materials and Building Structures, Delft, Holland, 54 p.

40. VOS, B. H., 1970: Moisture in monuments. Unpublished report of TNO for Building Materials and Building Structures, Delft, Holland.

41. VOÛTE, C., 1963: Some geological aspects of the conservation project for the Philae Temples in the Aswan Area. Geologische Rundschau, **52** (**2**), 665—675.

42. WELLMAN, H. W., and A. T. WILSON, 1963: Saltweathering, a neglected geological erosive agent in coastal and arid environments. Nature, **205** (**4976**), 1097—1098.

43. WELLMAN, H. W., and A. T. WILSON, 1968: Saltweathering or fretting. Encyclopedia of Geomorphology, Vol. III (E. W. FAIRBRIDGE, ed.), pp. 968—970.

44. WINKLER, E. M., 1965: Weathering rates as exemplified by Cleopatra's Needle in New York City. Journal Geol. Education, Vol **13** (**2**), 50—52.

45. WINKLER, E. M., and P. C. SINGER, 1970: Crystallization pressure of salts in stone and concrete. Manuscript in press. in: Geolog. Soc. America Bull., **83** (**11**).

46. WINKLER, E. M., E. J. WILHELM, 1970: Saltburst by hydration pressures in architectural stone in urban atmosphere. Geolog. Soc. America Bull., **81** (**2**), 567—572.

7. Chemical Weathering

The decay process of the natural earth material rock and stone has been well known to architects through many centuries. We do understand most phases of physical and chemical weathering as the interaction of the atmosphere with minerals and rocks, to obtain equilibrium conditions at the earth's surface. Much less is known about the time which these processes take. An attempt is here made to summarize the present knowledge of weathering and weathering rates applied to stone attack.

The quantitative approach to weathering was attempted by Sir ARCHIBALD GEIKIE in 1880, in his study of the weathering rates of headstones on Edinburgh graveyards. Work in this direction was continued by A. A. JULIEN (1884) on stone in New York City. Quantitative research on weathering rates was not resumed in Europe until after World War II, when extensive damage to buildings and monuments exposed weathered margins beneath soot-covered surfaces and revealed corroded carbonate rocks against undissolved reference points.

A few well-recorded cases give indications but they do not yet permit a generalization of weathering rates, because stone is a heterogeneous substance. Basalts from different localities are not exactly the same basalt. Sandstone have yet a larger variety, because the cementing mineral and the degree of cementation are to be considered. No attempt should be made to date buildings and monuments from the degree of stone decay unless a "corrosiveness index" of the atmosphere and natural waters has become available. In the following discussions of chemical weathering, distinction is made between true solution (or dissolution) without residue, and the dissociation of silicates by hydrolysis, with residual clay (solubilization).

7.1. Weathering by Solution

1. Solution is the complete dissociation of a mineral in a solvent, such as water. The solubility of a mineral in water is the mass of substance contained in a solution which is in equilibrium with an excess of the substance, given here in mg. per liter or parts per million. The solubility data are limited here to the common minerals gypsum, calcite, dolomite and others; minerals with great solubility, such as chlorides and sulfates, are given in g. per liter. The chemist expresses solubility as the solubility product (ion product constant) which is the concentration of ions in the saturated solution of a difficulty soluble salt, in moles per liter of solution.

The mineral substance tends to be attacked by the solvent till saturation is reached. Solution attack progresses more violently the less saturated is the solution.

The solution rate generally depends on the solubility of the salt, on the degree of original saturation of the solvent, and on solvent motion which may keep the solvent undersaturated. A condition of permanent undersaturation applies to the contact of stone with rainwater or streamwater on a building or on bridges and jetties below water level.

Solvent motion: KAYE (1957) determined the influence of solvent motion by experiments using dilute HCl on marble. He found that an exponential increase of the solvent flow results in a linear widening of the solution channel (Fig. 114). In the experiment, 2 cm.^{3}/sec flow velocity widened the channel diameter 0.4 mm., whereas 10 cm.^{3}/sec flow about 1.0 mm., and 40 cm.^{3}/sec flow only 1.6 mm. These values may well apply to natural conditions where the doubling of the flow velocity of corrosive river water in contact with jetty stone fortunately does not double the surface corrosion.

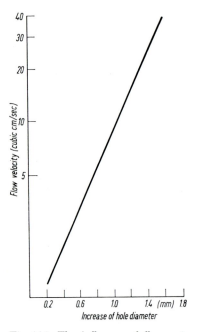

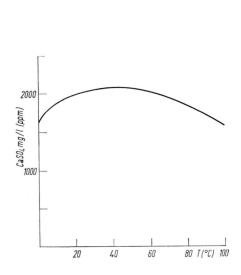

Fig. 114. The influence of flow rate on the rate of solution of limestones by dilute HCl. From KAYE (1967)

Fig. 115. Solubility of gypsum in water as a function of temperature

The solution of the mineral substance is influenced either by temperature only, as with gypsum, or by temperature and the pH of the solvent, as with the carbonates.

a) Solubility by temperature only: most salts increase their solubility with increasing temperature of the solvent, like the very soluble halite and some other chlorides, the magnesium sulfates, etc. The solubility of gypsum for various temperatures is given in Fig. 115. A maximum in the solubility lies near

$40°$ C with almost 2100 ppm (2.1 g./l.), but at temperatures near freezing only 1600 ppm and near $100°$ C 1575 ppm. KIESLINGER (1962) observed the magnitude of solution of underground waters in gypsiferous rocks and concludes that gypsum is about 32 times more soluble than calcite under average natural conditions. The solution attack of gypsum introduces large quantities of the sulfate ion into the solvent. A high sulfate content in natural waters endangers concrete, because the sulfate ion readily and rapidly attacks the alkalic portland cement resulting in crumbling of concrete. Gypsiferous rocks, including alabaster, should not be used on exterior walls, whether polished or unpolished.

Table 29. *Solubilities of Some Important Water Soluble Minerals, in* ppm *and the Solubility Products*

Mineral	Freshwater (ppm)	(solubility product)	Seawater (ppm)
Carbonates:			
calcite ($CaCO_3$)	$40-85$	$(0.99 - 0.87 \times 10^{-8})$	66
dolomite (Ca, Mg [CO_3])	less sol.	$(3.96 \times 10^{-11}$(a)$10°$ C)	50
siderite ($FeCO_3$)	$10-25$	$(3.96 \times 10^{-11}$(a)$10°$ C)	—
Sulfates:			
gypsum ($CaSO_4 \cdot 2\,H_2O$)	2000	(1.95×10^{-4})	6000
anhydrite ($CaSO_4$)	less		$5000-6000$
epsomite ($MgSO_4 \times 7\,H_2O$)	262 g./l. ($20°$ C)		
	335 g./l. ($50°$ C)		
hexahydrite ($MgSO_4 \cdot 6\,H_2O$)	308 g./l. ($20°$ C)		
	355 g./l. ($60°$ C)		
thenardite ($NaSO_4$)	388 g./l. ($40°$ C)		
	453 g./l. ($60°$ C)		
Chlorides:			
bischofite ($MgCl_2 \cdot 6\,H_2O$)	2635 g./l. ($20°$ C)		
	2710 g./l. ($60°$ C)		
halite	264 g./l. ($20°$ C)		
	274 g./l. ($60°$ C)		

b) Solubility by the presence of the CO_2 anion: The solubility curve for CO_2 in pure water is given in the chapter on Weathering Agents Fig. 79. All carbonate minerals are influenced by the amount of CO_2 present in the solvent. The presence of CO_2 in the water as carbonic acid accelerates the solution process, because solid calcite is converted to the very soluble Ca-bicarbonate, in the following manner:

1. $CaCO_3$ (solid) $\rightarrow CaCO_3$ (solution),
2. $H_2O + CO_2 \rightarrow H_2CO_3$; $H_2CO_3 + CaCO_3 \rightarrow Ca(HCO_3)_2$ (calcium bicarbonate).

A workable calcite saturation curve was first calculated and plotted by BAYLIS in 1935 for the purpose of quick identification of the degree of corrosion in water pipes, on the one hand, and scale formation on the other, based on the pH and the $CaCO_3$ already in solution. THRAILKILL (1968) extended the simple Baylis curve

into a more elaborate three-dimensional plot on a double log scale. Fig. 116
increases the legibility of the original diagram by adding the CO_2 in per cent, and
the $CaCO_3$ content in ppm. The pH of a solvent, i.e., the acidity, not only reflects
the CO_2 content but also the presence of other acids. Saturation lines are drawn

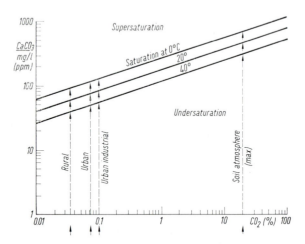

Fig. 116. Saturation diagram for calcite in CO_2-rich waters. Saturation equilibria are drawn
for rural, urban and urban industrial rain after WINKLER (1966), for soil atmosphere from
YAMAGUCHI (1967). The graph is modified from THRAILKILL (1968)

Fig. 117. Damage to limestone-marble by solution, marked by general retreat of the original
stone surface. Cracks originally invisible are now deepened by solution and widened by gashes.
Renaissance fountain, Venice, Italy

for 0°, 20°, and 40° C; the lines connect all points at which saturation occurs: the
solvent is then in equilibrium with calcite and solution stops. The approximate
solubility of rainwater may be scaled off readily along the dashed lines for rural

areas with 0.034% CO_2, for urban areas with about twice the quantity of CO_2, and for urban industrial about three times the CO_2 of rural environments. For comparison, the CO_2 content of the soil atmosphere in a sandy loam (17%) is also plotted. Rural rainwater can dissolve 95 ppm of calcite at $0°$ C, 120 ppm in an urban atmosphere (Fig. 117), 140 ppm in industrial urban, but almost 700 ppm in soil with 17% CO_2; marble buried in such soil would be exposed to considerable attack. The degree of undersaturation equals the corrosiveness of the water for carbonate rocks provided that solvent movement prevents saturation. Carbonate solution in tropical humid climate records much higher solution rates than in moderate humid areas. This observation seems to conflict with the solubility curves. Here, chemical solubility appears to be superimposed upon the activity of abundant chelating microfauna and microflora. A few particular aqueous environments are discussed in the following:

Stream waters: Bank fortifications, dams and locks are exposed to chemical attack along the shores of a stream. Attack also occurs to silt and clay in suspension, sand and gravel in the stream bed and rock exposed to, and contacted by water. These sources contribute to the supply of ions and toward an increase of the water hardness and reduced corrosiveness of the water. Ingredients are also added by atmospheric precipitation and through groundwater that feeds into the streams. Soft waters tend to reach equilibrium condition by chemical attack with the rocks they contact; the softer the waters the stronger the corrosive action. TUTHILL (1956) describes extensive damage to concrete by soft, lime-hungry waters within only a few years of exposure. Streams derived from arid or semi-arid regions are often loaded with chloride, sulfate and alkalinity. The effect of aggressive ions is discussed in the chapter on Weathering Agents.

Lake waters: Large lakes show little fluctuations in water level and chemical composition. Rainwater, dry fallout, leaching of ions from rock along shores and lake bottom, and inflowing groundwater supply the ions in humid climates. Smaller lakes are exposed to frequent fluctuations. Soft-water lakes have a $CaCO_3$ concentration of less than 40 ppm and a pH of $6.8—7.4$; the water is always corrosive toward carbonate rock and concrete. Hard-water lakes are generally located within areas of glacial drift or carbonate rocks; their $CaCO_3$ content is generally $40—200$ ppm and of a pH of $8.0—8.7$. These lakes are generally lime-saturated and are not aggressive. More information on the ion concentration of natural waters can be found with GORHAM (1961). The Great Lakes offer a good example for the influence of the rock environment and time of exposure. Lake Superior, the highest lake above sea level and mostly surrounded by igneous rocks is the softest and most corrosive of the Great Lakes, Lake Ontario is the lowest and the least corrosive, surrounded by carbonate rocks and glacial drift.

Seawater can be corrosive to rocks and metals, with a total of 34.400 ppm salts. Despite theoretical saturation of seawater with $CaCO_3$, beachrock erosion is well known in calcareous rock used on seawalls, jetties, etc. REVELLE and EMERY (1957) observed strong solution only at the intertidal zone (between high and low tide), where strong seasonal and diurnal undersaturation can occur by the influence of CO_2 emission by algal photosynthesis. Cold seawater and strongly diluted harbor waters may also become locally undersaturated and quite aggressive. On the other hand no corrosion on carbonates could be observed in many cases despite strong undersaturation of seawater. Chave (personal communication, 1966) found that everpresent hydrocarbons tightly coat sand grains and rock surfaces with a very thin organic (monomolecular?) layer protecting the inorganic mineral substance

Fig. 118. Marble headstone: darker elevated quartz-pyrite veins show relief against strongly reduced marble surface. South Bend, Indiana

Fig. 119. Differential solution along cracks and joints in Indiana limestone. Solution has been greater along the more porous shell bed (marked by arrows). Quarry face is smoothed by employment of channeling machines. Bedford, Indiana

from further chemical attack by seawater. Only grains with a damaged organic skin become vulnerable to solution attack. This protective coating may be readily destroyed by strong wave impact or foraging organisms.

Groundwater: Stone and concrete aggregate encounter occasional contact with groundwater in deep foundations. The slow flow of groundwater through rock pores and narrow channels gives the water enough time to reach near equilibrium in carbonate rocks. The ion content of groundwater is subject to extreme variation depending on both the source rock and the length of travel of the water in its medium. Glacial drift and limestone areas provide enough $CaCO_3$ that these waters are saturated and not corrosive to carbonate rocks nor to concrete any longer.

Differential solubilities: Building stones and head stones of grave yards are often surface-sculpted with progressive exposure to solution. Quartz, pyrite, chert nodules and mica flakes are insoluble components of many carbonate rocks. Chert in limestone and quartz veins in marbles show up distinctly (Fig. 118). Mica in some marbles form sharp blades freed from the surrounding calcite matrix by solution. Less easily understood is the formation of reliefs by solution in calcite against a calcite matrix, as is the case with many fossiliferous limestones and limestone conglomerates (Fig. 119). The fossiliferous Gizeh limestone of Egypt used as the pedestal for the Cleopatra's Needle monument in New York was

Fig. 120. Strong surface relief of fossiliferous limestone by differential solution in an urban atmosphere. Irregular horizontal crack, roughly parallel to bedding plane, has been widened by solution; cracking is also possible by salt action. Gizeh limestone from Egypt, with nummulites (torpedo-shaped) standing out clearly. Pedestal to Cleopatra's Needle, New York

exposed to differential solution for only 90 years. Solution of the matrix enclosing the disc-shaped nummulites forms the rough surface which appears to be further enhanced on the photo by long shadows of the setting sun (Fig. 120). Differential solution can also be seen in some limestones along stylolites; moisture penetrates readily along stylolitic zig-zag clay seams forming V-shaped notches and valleys. Figs. 121 and 122 describe the formation of solution damage along stylolites in urban New York City and on a shower stall.

Solubility rates (surface reduction rates): Reduction rates of limestone and marble surfaces are sparse and can therefore be only approximated for rural and suburban

Fig. 121. Weathering of stylolites in limestone: Weathering progresses deeply along clay fillings. Rainwater in urban polluted atmosphere quickly widens the opening by solution. Statue of lion in front of Public Library, New York. Limestone sculpture sits on granite-gneiss

Fig. 122. Stylolites in Holston limestone-marble, deeply gouged by solution along the vertically set stone. The stone, unfortunately, is very popular for showerstalls, where damage occurs by moisture running down the marble wall. The depth of corrosion along clay seams reaches $\frac{1}{2}$ inch. Corrosion appears to be concentrated within 2 feet above the floor of the shower stalls. The marble was installed about 35 years ago

atmospheres in Fig. 123. No information was available for urban industrial atmospheres. Approximate surface reduction may be scaled off from the graph for various rainfall averages. The following well documented cases were used in the Figure:

a) A fine-grained Vermont marble from a suburban South Bend grave yard. The photo of Fig. 118 shows the strong relief, an average rate of $1\frac{1}{2}$ mm. was recorded in about 50 years for 35″ precipitation for the South Bend area. The measurements were made against an unreduced quartz-pyrite vein with a light green tarnish on the pyrite. The original contours of the stone surface are still evident on top of the veins. The marble solubility is much greater within the quartz-pyrite veins than further away because sulfuric acid from the pyrite veins adds corrosiveness to the existing carbonic acid from the atmosphere. The highest solubility is believed to occur in spring and late fall when near-freezing rainwater

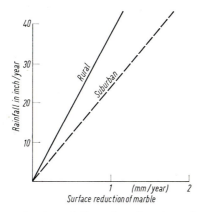

Fig. 123. Surface reduction rates by solution of carbonate rocks. After WINKLER (1966)

temperatures are associated with CO_2- and SO_4-ion emission from heating of homes and office buildings; maximum acid emission combines with maximum solubility of CO_2 in waters at low temperatures.

b) Unspecified limestone in New York, after LOUGHLIN (1931); at 43 inch precipitation, the reduction rate is 1.57 mm. in 50 years. Increased industrial activity today would rapidly increase these figures.

c) Theoretical calculations of limestone solubility after MILLER (1952) with 40-inch annual precipitation in a rural environment result in 0.4 inch in 450 years (this corresponds to about 1 mm. in 50 years).

All three data fit approximately the curves in Fig. 123 for rural and suburban atmospheres. Texture, structure, and impurities, however, influence the solution rates of carbonate rocks to a large extent.

7.2. Weathering of Silicate Rocks

Silicate minerals consist of more or less tight crystal lattices into which the metal cations Ca, Na, K, Mg, and others are built in. Leaching (solubilization) can unlock and free the ions to either serve as valuable plant nutrients or to be washed out and carried away by groundwater to streams to reach the ocean, the place of final destination. This process is fast if we consider it to be a geological process, but often very slow in terms of human life. The weathering process of silicate minerals is still not fully understood. LOUGHNAN (1969) reports the following possible processes involved in silicate weathering: Ions exposed at the surfaces possess unsaturated valencies which hydrate in contact with water. The very small H^+ ions with their high charge and very small ionic radius readily penetrate the mineral surface and break down the silicate structure. The following apparently simultaneous processes appear to take place:

1. The parent mineral breaks down with the release of cations and silica, whereby silica in the silicate may either retain its original atomic arrangement or enter the dispersed state.

2. Metal alkalies are removed by solubilization as a result of the breakdown.

3. Reconstitution of the residue may form new minerals from components of the atmosphere which are in a stable or metastable equilibrium with the present environment. The entry of the water molecules takes place into spaces left vacant after leached metal cations (hydrolysis). The process leads to the formation of clays. Kaolinite appears to be the most important clay mineral product of stone weathering when much rain is present with intense leaching and removal of Ca, Mg, Fe, Na, K and some silica under oxidizing conditions (KELLER, 1956). The mechanism of kaolinization is sketched in Fig. 124. The rate of leaching enables us to

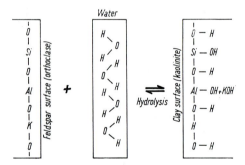

Fig. 124. Mechnism of feldspar weathering by hydrolysis: from orthoclase to kaolinite. From LOUGHNAN (1969)

set up weathering models which were designed for watersheds based on the transport of ions by the streams. The leaching rates of stone above the ground is very different in character, because no interaction takes place between the biologically and chemically active soil with the unweathered rock. The loss of ions in igneous rocks during incipient weathering is a maximum for calcium and a minimum for aluminum. The ionic mobility is summarized by LOUGHNAN (1969) in Table 30.

BOROVEC and NEUŽIL (1966) found that, of the feldspars, the Na-plagioclase albite leached fastest, the K-feldspar orthoclase slowest. KELLER et al. (1963) tried to quantify the weathering of silicate minerals by leaching common rock-forming silicates with both distilled and with CO_2-charged water. CO_2-charged water had doubled the leaching rates for Ca, Mg, Na, and K. Orthoclase and muscovite are much more sensitive to high CO_2 content than other minerals. Quartz, the most stable form of silica, dissolves at a rate of only 4—5 ppm. A variety of powdered igneous rocks was exposed to leaching in distilled water by KELLER and REESMAN (1963): minerals and rocks with a low silica content were found to lose silica more easily than minerals and rocks with a higher silica content. In stone

Table 30. *Relative Mobility of Common Ions*

Ions	Mode of Leaching
1. Ca^{2+}, Mg^{2+}, Na^+	leach readily under favorable conditions
2. K^+	leaches readily; can be retarded by fixation of potassium in silicate lattice to form illite
3. Fe^{++}	leaching rate depends on redox potential
4. Si^{4+}	slow loss under leaching conditions; accelerates at $pH = 10$
5. Fe^{+++}	immobile under oxidizing conditions, mobile if reducing
6. Al^3	immobile at pH 4.0—9.5; very soluble below and above

After: LOUGHNAN (1969).

weathering above the ground surface, DENNEN and ANDERSON (1962) observed that temperature and relative humidity of the immediate environment play a more important role than the actual mineral composition of the rock. The oxidation of the ferrous iron is the earliest evidence of incipient weathering, followed by the removal of Na, Ca, and Mg. It would be impossible to discuss the great volume of literature on stone weathering here; only a few of the most important references have been mentioned.

Degree of weathering: The weathering sequence of silicate minerals leads from a fresh granite to a sandy clay. The degree of kaolinization strongly influences both the rock strength and its durability. The clay content of a rock in the process of weathering is difficult to quantify by standard analytical mineralogical methods. GRANT (1969) approximates the clay content by the "abrasion pH" of the weathered rock which is determined by the pH of slurried fine-ground rock substance. Low pH values are brought forth by introduced acid H^+ ions essential in the weathering of silicate minerals. Fig. 125 approximates the clay content of a granite from the measured abrasion pH. The SiO_2-to-Al_2O_3 proportion is another weathering index which could indicate the degree of stone soundness. The exact proportion of the fresh rock would have to be known.

Rates of silicate weathering are extremely difficult to evaluate because too many factors are involved which may influence the process. Stone exposed to polluted atmosphere may show early signs of incipient weathering by undesirable discoloring, loss of luster and hardness of feldspars and ferro-magnesian silicates. STRAKHOV (1967) estimates an increase of the weathering rate by solubilization and hydrolysis 2 to 2.5 times with each temperature increase of 10° C, about 20 to 40 times more for tropical moist areas than in moderate climate with less precipitation. The rate of silica leaching by percolating moisture may serve as a scale of weathering of silicate rocks as the SiO_2-to-Al_2O_3 ratio (RUXTON, 1968) as mentioned before in the paragraph Degree of Weathering. STRAKHOV assumes a leaching rate of 10 ppm of silica to reach 27% silica from a rock which originally contained 47%; thus 300 years would be needed in a tropical moist climate with 3750 mm. precipitation to remove this quantity of silica, but 2200 to 4400 years in a temperate climate with 250—500 mm. STRAKHOV's calculations are, no doubt, of interest to the stone industry.

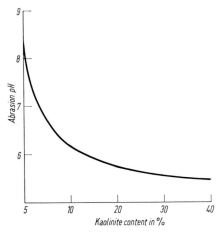

Fig. 125. Degree of granite weathering based on the abrasion pH versus the clay content of the granite. From GRANT (1969)

ČERNOHOUC and ŠOLC (1966) calculated weathering rates of basalts on mediaeval castles in Bohemia: The recorded penetration by weathering with time is approximately exponential (Fig. 126). The oxidation of siderite to ferric

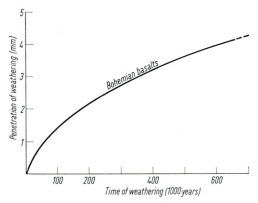

Fig. 126. Approximate weathering rate of Bohemian basalts. Weathering rate appears to slow down exponentially. From ČERNOHOUC and ŠOLC (1966)

oxide is similarly exponential as seen in Fig. 127. After about three years of exposure, half of the siderite powder had oxidized to hematite, and in 35 years all the siderite was oxidized.

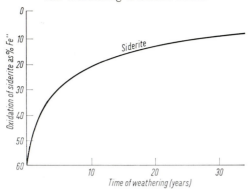

Fig. 127. Weathering rate of siderite ($FeCO_3$) to hematite (Fe_2O_3). Note similarity of curve with weathering curve of basalt. Oxidation rate data from SCHALLER and VLISIDIS (1959)

Fig. 128. Rapid decay leading to total destruction of relief on tympanon of parish church in Opherdicke (near Unna, Westfalia). Destruction was severe in 1893, after 600 years of exposure; damage was catastrophic in the last 80 years. Soest greensandstone (Upper Cretaceous, 64% calcite, 18% glauconite, 18% quartz). Photos and information supplied by Dr. SCHMIDT-THOMSEN, Landesdenkmalamt Westfalen-Lippe, Germany

Some basalts have weathered anomalously fast, developing grey spots on the black basalt surface, the first sign of rapid deterioration. Crumbling may follow within about 5 years of exposure. This phenomenon has been called "sunburned basalt" (Sonnenbrennerbasalt); the name supports the old belief of surface burns by solar radiation followed by rapid crumbling of the rock. ERNST (1960) ascribes this phenomenon to a finely crystalline network of light-grey concentrations of analcite along which fine cracks can develop; these are believed to be the possible source for the rapid decay. Basalt as building material should be carefully tested both mineralogically and for cracks and microcracks. Boiling the rock in concentrated potassium hydroxide shows discoloring, and treatment with ethylene glycol leads to disintegration in a short time in the laboratory.

Fig. 129. Rapid decay of statue in garden of castle Overhagen near Lippstadt. The statue was set up in the first part of the 18th century. Little damage in photo of 1912, but statue is almost destroyed in recent picture. Porous Baumberg sandstone (Upper Cretaceous, 50—80% calcitic matter, 10—30% quartz, etc.) is located near the edge of industrial Rhein-Ruhr area. Photos and information were supplied by Dr. SCHMIDT-THOMSEN, Landesdenkmalamt Westfalen-Lippe, Germany

This author believes that the weathering rate of silicate rocks may be exponential; the exponential character may be reversed, very slow in the beginning when the rock is still quarry-fresh but progressing rapidly once the lattice destruction has started; only weathered rock surfaces harbor acid-secreting and chelating bacteria and fungi. It is believed that both cases can take place, possibly combined.

The Bohemian basalts may have been slightly weathered when the mediaeval builders took the stone from the quarries; columnar jointing accelerated quarrying operations, but also granted access to weathering. Weathering penetrated quickly after the first centuries of exposure, but slowed down as the weathered layer may have given some protection to the unweathered stone substance beneath.

The existence of a corrosiveness index would be a help in the calculation of weathering rates, but only a drop in the bucket of variables which strongly influence weathering rates. A prediction of the exact rate of attack on a particular building or monument may therefore never be possible.

Fig. 130. Strongly weathered and surface reduced masonry wall of Soest greensandstone (Upper Cretaceous), composed of 64% calcite, 18% glauconite, 18% quartz. St. Walburga church, Werl near Soest, Westfalia. The church was built in the 14th century. The sandstone was strongly attacked by the humid climate of the industrial Rhein-Ruhr area of Germany. Photo and information were supplied by Dr. Schmidt-Thomsen, Landesdenkmalamt Westfalen-Lippe, Germany

A comparison of the visual rate increase of monument decay is given in the two photos of a sandstone sculpture at Herten castle near Recklinghausen Fig. 72a, b, Westfalia, located at the beginning of this book; the visible degree of stone decay is plotted on a graph facing the pictures. Other remarkable time-lapse photos of monument decay are presented in Figs. 128a, b and 129a, b. Fig. 130 is an excellent example of surface reduction in layered sandstone.

Summary

Chemical weathering is the most powerful tool in the destruction of stone. It may occur essentially as solution, as with carbonates and sulfates, and solubilization with silicates, accompanied by recrystallization of the original mineral towards a more stable form. The process consists of leaching and hydration.

Solution: Carbonates, gypsum and salt rock are soluble without residue. Solution rates depend on temperature of the solvent, flow velocity and on the CO_2 content for carbonates. High CO_2 contents of urban atmospheres can double and triple the weathering (solution) rates of limestone, dolomite and marble.

Solubilization of ions from silicate minerals and rocks frees ions for further transport away and exposes the gaps left in the crystal lattices to hydration, the entry of the OH ion. Ca, Mg, Na are leached faster than Al. Discoloring and softening of margins is followed by pitting and crumbling of the stone surface.

In general, weathering rates of silicate rocks — except for some basalts — progress slower than with carbonates.

References

1. Borovec, Z., and J. Neužil, 1966: Experimental weathering of feldspars by hot water. Acta Universitatis Carolinae, Geologica, **3**, 207—222.
2. Černohouc, J., and I. Šolc, 1966: Use of sandstone wanes and weathered basaltic crust in absolute chronology. Nature, **212 (5064)**, 806—807.
3. Dennen, W. H., and P. J. Anderson, 1962: Chemical changes in incipient rock weathering. Geol. Soc. America Bull., **73**, 375—384.
4. Ernst, Th., 1960: Probleme des „Sonnenbrandes" basaltischer Gesteine. Z. Deutsche Geolog. Gesellschaft, **112**, 178—182.
5. Geikie, A., 1880: Rock weathering as illustrated in Edinburgh church yards. Proc. of the Royal Soc. of Edinburgh, 1879/80, pp. 518—532.
6. Gorham, E., 1961: Factors influencing supply of major ions to inland waters with special reference to the atmosphere. Bull. Geol. Soc. America, **72 (2)**, 795—860.
7. Grant, W. H., 1969: Abrasion pH, an index of chemical weathering. Clays and Clay Minerals, **17 (3)**, 151—155.
8. Julien, A. A., 1884: The durability of building stones in New York City and vicinity. U.S. 10th Census, 1880, vol X, Special Rept. on Petroleum, Coke, Building Stone, Chapter V, pp. 364—384.
9. Kaye, C. A., 1957: The effect of solvent motion on limestone in solution. The J. of Geology, **65**, 35—46.
10. Keller, W. D., 1956: Clay minerals as influenced by environments of their formation. Bull. Am. Assoc. Petr. Geol., **40 (11)**, 2689—2710.
11. Keller, W. D., W. D. Balgoard, and A. L. Reesman, 1963: Dissolved products of artificially pulverized silicate minerals and rocks, Part I. Journal Sedimentary Petrology, **33 (1)**, 191—204.
12. Keller, W. D., and A. L. Reesman, 1963: Dissolved products of artificially pulverized silicate minerals and rocks, Part II. Journal Sedimentary Petrology, **33 (2)**, 426—437.
13. Kieslinger, A., 1962: Wasserstollen im Gipsgebirge. Österr. Ingenieur-Zeitschrift, **5**, 338—350.
14. Loughlin, G. F., 1931: Notes on the weathering of natural building stones. Am. Soc. Testing Materials, Proc., **3 (II)**, 759—767.

15. LOUGHNAN, F., 1969: Chemical weathering of silicate minerals. Elsevier Publishing Co., New York, 154 p.

16. MILLER, J. P., 1952: A portion of the system $CaCo_3 - CO_2 - H_2O$. Am. J. Science, **250**, 161—203.

17. REVELLE, R., and K. O. EMERY, 1957: Chemical erosion of beach rock and exposed reef rock. U.S. Geol. Survey Prof. Paper 260-T, pp. 699—709.

18. RUXTON, B. P., 1968: Measures of the degree of chemical weathering of rocks. Journal Geology, **76 (5)**, 518—527.

19. SCHALLER, W. T., and A. C. VLISIDIS, 1959: Spontaneous oxidation of a sample of powdered siderite. American Mineralogist, **44**, 433—435.

20. STRAKHOV, N. M., 1967: Principles of lithogenesis. Oliver & Boyd, Edinburgh (translation), **I** (245).

21. THRAILKILL, J., 1968: Chemical and hydrologic factors in the excavation of limestone caves. Geol. Soc. America Bull., **79 (1)**, 19—46.

22. TUTHILL, L. H., 1956: Hardened concrete; resistance to chemical attack. Am. Soc. Testing Materials, Spec. Techn. Publ., **169**, 188—200.

23. WINKLER, E. M., 1966: Important agents of weathering for building and monumental stone. Engineering Geology, **1 (5)**, 381—400.

24. YAMAGUCHI, M., W. J. FLOCKER, and F. D. HOWARD, 1967: Soil atmosphere as influenced by temperature and moisture. Soil Science Soc. America Proceedings, **31**, 164—167.

8. Stone Decay by Plants and Animals

Plants and animals can attack stone to a large extent both by mechanical and by chemical action. While the action of higher plants may be both mechanical and chemical, bacteria, the lowest kind of life, attack by chemical means only. The biotic decay is very complex and this topic has been deplorably neglected in the past.

8.1. Bacterial Activity in Stone Decay

The basic types of bacteria should be distinguished here and their activity sharply separated, the autotrophic and the heterotrophic bacteria. Autotrophic bacteria take their energy from the sunlight, chemical oxidation and reduction, whereas heterotrophic bacteria obtain their energy from existing organic substances of the soil.

a) *Autotrophic bacteria* play an important role in incipient weathering of minerals and rocks. According to their activity, we distinguish:

Nitrogen-fixing bacteria: The presence of calcium nitrate $Ca(NO_3)_2$ on limestone walls was observed and discussed by KAUFFMANN (1960). The chemical process by nitrificating bacteria appears to be as follows:

$$2\,CaCO_3 + (NH_4)_2SO_4 + 4\,O_2 = Ca(NO_3)_2 + CaSO_4 + 2\,CO_2 + 4\,H_2O.$$

KAUFFMANN therefore concludes that biologic degradation of carbonate rocks is primarily based on nitrification. Some gypsum results from nitrification; the great bulk of gypsum on carbonate surfaces, however, is usually the product of the direct interaction of sulfates of the atmosphere with carbonate rocks.

Sulfur bacteria: Thiobacillus thiooxidans and Th. thioparus are the most common of the sulfur bacteria; they can oxidize sulfur by providing an acid oxidizing solution effective in stone attack. Thiobacillus was found by MILLOT and COGNE (1967), several thousand per gram of rock, in the sandstone of the cathedral of Strasbourg. Thiobacillus was credited with converting sulfur from bird droppings to sulfite and further to sulfate; small crystals of gypsum were found between the sand grains, provided by the 2% of calcite in the sandstone cement. Desulfofibrio desulfuricans can form the little-soluble $CaSO_3$ by reduction; the solubility is not influenced by carbonic acid in the solvent. Sulfuric acid, on the other hand, forms gypsum, which is much more water soluble than $CaSO_3$. POCHON (1968) estimates a maximum number of 1 million bacteria per gram of stone substance on stone surfaces. Sulfur-oxidizing and nitrifying bacteria are relatively small in number. In the rain forest of tropical Java 16 million bacteria per gram of rock were counted (VOÛTE, 1969). Sound stone is not readily invaded by bacteria; fungi start their activity during the beginning of weathering and on these, other organisms find a foothold (WEBLEY *et al.*, 1963). Bacteria can break up silicate minerals as quickly as fungi do. Weathered rock develops a large bacterial population along all surfaces as well as along cracks. A clean, truly fresh rock surface lacking a bacterial population should be expected to have a longer life span than rock with an initial microbial population. The population on the volcanic rock of Borobudur appears to be exceptionally high. No figures are available on weathering rates in nature which were influenced by lichens and bacteria as compared to a stone surface which was kept continuously clean and sterile. Silicate rocks with traces of incipient weathering should be avoided, because decay can be expected to progress rapidly once it has got a start.

Iron bacteria, as the autotrophic Ferrobacillus ferrooxidans and Thiobacillus ferrooxidans: These bacteria are the most important oxidizers of iron. In the oxidation of pyrite, sulfur bacteria also play an important role. SILVERMAN and

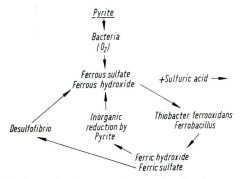

Fig. 131. Oxidation and reduction of iron and sulfur by bacterial action.
Modified from SILVERMAN and EHRLICH (1964)

EHRLICH (1964) sketched a flow chart of chemical reactions as they are induced by both iron and sulfur bacteria. Fig. 131 presents a modification of the original chart.

Calcite bacteria, Arthrobacter, was found on limestone surfaces by DRZAL and SMYK (1968) where they participate in the dissociation of carbonate rocks.

b) *Heterotrophic bacteria* live off available organic substance which has been produced by higher organisms, such as lichens, on stone surfaces. Heterotrophic bacteria have only limited effect on stone and are much less important in stone decay than the autotrophic fauna.

8.2. Algae, Fungi, Lichens, and Mosses

DRZAL and SMYK (1968) studied their effect in the laboratory with bacteria. Actinomycetes and fungi attack silicate minerals, especially the micas, orthoclase and others, by the production of carbonic, nitric, sulfuric and some weaker acids. Most recently, SILVERMAN and MUÑOZ (1970) studied the activity of the common fungus Penicillium simplicissimum WB-28 in vitro. Significant amounts of Si, Al,

Fe, and Mg were freed by the secretion of nitric acid, whereby the pH of the initial solution of 6.8 dropped to 3.5 in only 7 days. The degree of solubilization (leaching) of Si ranged from 0.3—31%, of Al 0.7—11.8%, but of Fe 25—60%. In general, basic igneous rocks were much more easily attacked than acid granitic rocks, which apparently remain very resistant to fungal attack. Iron and magnesium record the highest loss in all rocks tested, whereas Si and Al were only removed from basic igneous rocks. Urban and industrial locations tend to rid themselves of lichens and mosses as these plants are rejected by the presence of soot or sulfate. SCHAFFER (1932) describes damage to stone by lower plants in considerable detail. The symbiotic nature of lichens, algae and fungi is characterized by slow growth and the ability to attach to bare rock surfaces without the need of supporting soil; they are the pacemakers in the formation of humus, which in turn supports higher plants later. SCHAFFER distinguishes between calcicolous lichens which prefer carbonate rocks as substrata and silicolous lichens which prefer silicate rocks. The lichens live either within the rocks or on their surfaces. Lichens growing within

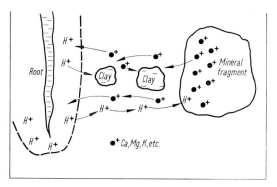

Fig. 132. Bio-chemical interaction of a plant rootlet with a silicate mineral fragment.
From KELLER (1957)

aggregates of translucent minerals bring forth discolored patches of brown and pink. Silicolous lichens are primarily found on igneous rocks on top of the mineral surfaces only. Stone surfaces are attacked by lichens, as follows:

a) By mechanical water retention: the rough surface and spongy character of the lichens retain water for a long time and keep the stone surface moist underneath; this may contribute to damage later as the stone cannot rid itself of moisture.

b) By ion exchange and secretion of acids: KELLER and FREDERICKSON (1952) describe the weathering process by plant root action in much detail. The very small H^+ cation produced by the rhizoms of lichens as well as by the root systems of higher plants and climbing vines such as Boston Ivy, Virginia Creeper, and others, easily exchange with negatively charged nutrient metal cations with the minerals and soils; this exchange is also important in the break-up of the CO_3 anion of the carbonates. Fig. 132 pictures the mechanism of ionic interaction of plant root systems with soil particles. Once the break-up process has started it is strongly accelerated by the action of carbonic, humic and various other complex organic acids which are produced from the organic remains. As soon as a film of colloidal clay has formed on the mineral surface, the acid-reacting H^+ continues its attack on the still-unweathered mineral substance beneath.

Table 31. *Microbial Analysis of a Soil Sample and Two Stone Pieces from the Temple of Borobudur, Indonesia*

Microfauna	Soil Sample A-Horizon	Stone Sample near Ground	Stone Sample 1 mm. above Ground
Bacteria, total	130,000	16,000,000	250,000
Fungi	3,000	230,000	13,000
Actinomycetes	100,000	3,000	30,000
Algae	6,000	90,000	6,000
Nitrogen-bacteria	1,475	12,500	12,100
Sulfur-bacteria	183	none	none

After: VOÛTE (1969). All figures are individuals per gram of rock.

Table 32. *Summary of Microbial Life and Its Activity on Stone*

Elements affected		Micro-Organism	Geological Activity
Ca, Mg		Arthrobacter	dissociation of calcite, dolomite
		Thiobacteria	only Mg attacked by bacteria
		Desulfofibrio	only Mg attacked by bacteria
		algae	only Mg attacked by bacteria
Fe	oxidation:	Thiobacter ferrooxidans	oxidizes ferrous ions
		Ferrobacillus	oxidizes ferrous ions
		algae	oxidizes ferrous ions
	reduction:	Desulfofibrio	reduces ferric to ferrous ions
		Clostridium	
N		Azobacter and others	dissociates calcite to Ca-nitrate, gypsum ("Mondmilch"?)
P		Penicillium and others	decomposes Ca-phosphate
S	oxidation:	Thiobacteria	sulfur to sulfates from
		fungi, filamentous	atmosphere on buildings
	reduction:	Desulfofibrio	reduction of sulfates to H_2S at
		yeast	or near ground level
Si		bacteria, fungi	dissolution of silicates
		Actinomycetes	
		Aspergillum	
		Penicillium	

The chart is modified by the author, after SILVERMAN and EHRLICH (1964).

Quantitative laboratory experiments with fungi and lichens on minerals and rocks were performed by HENDERSON and DUFF (1963). The fungi Aspergillum niger, Spicaria sp., and Penicillium sp. (type C_5 and C_2) produced considerable quantities of citric and oxalic acid from a 4% glucose solution; these acids are also believed to be most active on stone. The important ions Al, Mg, and Si were extracted from a few rock forming minerals under the influence of fungus culture solutions: the most active fungus appears to be Aspergillum niger, a common fungus of certain black lichens which removed 14% of the SiO_2 from biotite. The weathering rate of silicate rocks can therefore be expected to be accelerated under a cover of this common fungus. Some marine algae can do active boring within the rock substance by action of their terminal cells of endolithic filaments. The advance prefers cleavage planes and grain boundaries (GOLUBIC, 1969).

The microbial population is generally high in moderate semihumid and humid climate, but is much higher in humid tropical regions. VOÛTE (1969) analyzed weathered volcanic stone of the ancient Buddhist temple of Borobudur in Java, Indonesia. The results are summarized in Table 31.

The greatest microbiological activity in stone appears to be found near the ground above the soil level.

A summary of microbial activity is given in Table 32, which presents the bacteria and other micro-organisms actively involved in the formation or decomposition of stone.

8.3. Weathering by Biochemical Chelation

The dissociation of calcite and silicate minerals by other agents than the acidity brought forth by the presence of carbonic acid and organic acids was noticed by soil scientists some time ago. KELLER (1957) showed with a simple experiment that calcite dissociates in an aqueous solution of sodium salt of EDTA (ethylene diamine tetra-acetic acid) by chelation, during which reaction calcium is taken up in solution to remain in solution despite pH values of 10 and above, where no more calcite attack should be expected. The chelation process is the takeup of a metal cation into a void space of an organic ring structure in which it is held tightly; removal from the mineral surface leads to the dissociation of the mineral. SCHATZ et al. (1954) conclude that "where acidity does not function in biological weathering, chemical degradation of rocks and minerals appears mediated by chelation". Acid strength is not an index to the chelating ability. Chelation plays an essential role in the breakup of bare rock populated by primitive rock lichens under generally nonacid conditions as lichens consume CO_2 rather than produce it in metabolism. Lichens appear to be the most prolific producers of chelators, some of them replacing the function of soil. The authors exposed basalt, granite and some other rock types to organic chelators common in nature. Salycilate is the strongest, but tartrate, citrate, gluconate and others can chelate as well. Natural residual soils contain many chelating agents; these accelerate the breakup of rock material and provide nutrients to plants and crops in available form. The presence of lichens on surfaces of stone and concrete can thus be very harmful, as lichens and bacteria secrete chelators in abundance, and these dissociate rock and stone leading to solution or leaching under nonacid conditions.

8.4. Physical Action of Plant Roots

Besides the strong chemical action of plant roots, the roots of higher plants are known to have contributed to some minor mechanical breakup of stone. Grasses and small trees may find hold along crevices and mortar joints of older buildings and may thus expose their host to damage. Plant-root wedging along rock cracks is well known. GILL and BOLT (1955) measured both the axial and the radial root pressures of corn, beans and peas.

Fig. 133. Widening of natural joint in flat lying sandstone by growing juniper tree. Colorado National Monument, Colorado

Table 3?. *Axial, Radial Pressure of Roots of Corn, Beans, and Peas*

	Distance from Root Tip (in mm.)	Axial Pressure (in atm.)	Radial Pressure (in atm.)
Beans	0.0	19.36	6.11 (max)
	4.0	10.68	
	5.0	7.70	
Corn	2.6	24.9	6.95 (average)
Peas	2.2	13.33	—

After: GILL and BOLT (1955).

a) Axial pressures on root tips: pressures were measured along the long axis of the root; the maximum pressures were recorded along the root tips.

b) Radial pressures of roots are perpendicular to the long axis; the radial pressures are generally much smaller than the root tip pressures. The pressure becomes very effective as soon as axial growth becomes inhibited. The total root surface is actually much larger for radial than for axial pressures: the wedging force is therefore high and cracks and joints may be widened for easier access of moisture. TAYLOR and RATCLIFF (1969) correlate the root pressures with the plant osmotic pressures in growing cotton, peas and peanuts; the approximate root pressure is about 10% lower than the measured osmotic pressure of a plant. In Table 33 the axial and radial pressures of the bean, corn and pea are compared. A continuous average radial pressure of 15 atmospheres should be expected by advancing plant roots and rootlets.

STOLZY and BARLEY (1968) compared the root-tip pressures in soil with the resistance which a commercial soil penetrometer encounters in soil. A lower root-tip pressure of 6.1 kg./cm.^{2} compared with 7.3 kg./cm.^{2} penetration resistance of the soil is explained by the authors with the maximum efficiency and surface smoothness of the plant roots as well as their ability to find soil zones of minimum resistance. In general, a soil penetration by plant roots is not possible with a soil strength of 25—30 atmospheres, penetration is only occasionally possible with 20—25 atmospheres, and always possible with less than 19 atmospheres (TAYLOR and BURNETT, 1964). Based on these data, plant roots should not be expected to penetrate sound and uncracked rock or well-cemented mortar joints. Roots tend to grow along rough stone surfaces where they interact with the minerals chemically and extract sparse nutrients through surface attack while they continuously seek a weak joint or crack which will permit entry of the root tip (Fig. 133).

8.5. Boring Animals

Intertidal marine zones, which are the shore line between high and low tide, are most endangered by rock borers; this is the zone which may involve man-made sea walls of stone or concrete, or submerged historical monuments. The temple of Jupiter Seraphis on the Mediterranean shore at Puzzuoli near Naples, Italy, is marked by two distinct rows of densely spaced boring mussels, located today about 30 feet above the present sea level. It is evident that the temple was submerged twice into the intertidal zone within the last 1800 years.

Boring is achieved either by mechanical abrasion or by chemical etching through acid secretion. ANSELL (1969) describes the mechanical boring mechanism as follows: The strong foot of the animal can anchor the end of its foot near the exit of the hole against the wall by expanding its foot while the sharp-edged shell abrades the stone by turning back and forth around its long axis. Various species of angel wings (Pholas) and false angel wings (Petricola) are mechanical rock borers in carbonate rock, concrete and soft shale. Angel wings dig at a rate of 12 mm. a year, with a total depth of about 150 mm. WARME (1969) observed a higher population density of pholads in soft rock than in hard material. Sea urchins of the genus Echinus and Eucidaris are able to dig in by means of abrasion, sinking their sharp teeth, located in the center at the bottom of their body, into the rock underneath. Their rate of boring may be as much as 1 cm. a year in limestone, faster in softer rock.

Chemical borers, on the other hand, are restricted to carbonate rocks. Boring clams should be considered initial borers as these are followed by other boring organisms which dig an intricate network of channelways of various sizes leading to the final disintegration of the rock substance. By acid secretion, the clam, Lithophaga, digs straight, smooth channels to a depth of 100 mm. Boring mussles, Mytilus, are the most common of the boring higher animals in limestone. After the clams, Polychaete worms continue a network of small round or oval channels, generally densely spaced, and the sponge, Cliona, completes the work of fragmentation to a depth of about 25 mm., whereupon the friable debris crumbles as sand to the bottom. The primitive sponge, Cliona, bores by advancing plasma into the rock substance as "amoebocytes" whereby the main part of the cell with the cell nucleus remain in place after they have etched their outlines into the rock (COBB, 1969). Extensive bio-erosion in the subtidal zone of tropical Bermuda reaches about 1.4 cm. per year for Cliona, 1.3 cm. a year for the boring clam Lithophaga, whereas very little bio-erosion could be recorded in the intertidal zone by scraping browsers for algal growth; no bio-erosion was observed at all above the intertidal zone (NEUMANN, 1966). In general, bio-erosion is progressing at a sometimes catastrophic rate, often 25 times faster than ordinary marine erosion consisting of solution and mechanical abrasion by wave action.

8.6. Birds on Buildings

Urban buildings are often inhabited by a variety of birds seeking a place for rest, refuge and breeding underneath projections, pinnacles and other protected spots. Pigeons, sparrows and house martens are the most common and best adjusted to man's activities. Large quantities of excrements contain some phosphoric and nitric acids which etch stone or chemically react with carbonates to form calcium phosphates and some nitrates. Most of the sulfur in the surface coating of gypsum on the Strasbourg Cathedral is believed by MILLOT and COGNET (1967) to be derived from bird droppings. This author, however, rather assigns the sulfur to polluted air. A precise evaluation of the damage to stone by bird droppings is only possible under controlled laboratory conditions, because too many other variables interfere in nature.

Summary

Stone is readily attacked by a large variety of animals and plants. The attack observed appears to be mostly microbiological; higher plants and animals are also involved. The destructive work may be of biochemical or of mechanical nature.

1. Microbiological attack is mostly chemical, by chemical secretion and photosynthetic activity, producing and accelerating chemical reactions and secreting organic acids which readily attack silicate minerals, carbonates, and sometimes pyrite. Weathered rock surfaces may be populated by as many as one million bacteria per gram of rock, hundreds of thousands of fungi, and others, much more in tropical areas.

2. Higher plants interact with minerals, as the secreted small acid H^+ ion unlocks metal cations of the minerals. The destructive action of plant roots can be also mechanical by means of wedging within rock cracks and weak mortar joints. A radial root pressure of 15 atmospheres should be considered average.

3. Higher animals: Bird droppings on buildings help to corrode stone by organic acids. Marine animals may dig themselves into stone and concrete within the intertidal zone of the ocean shores of all climates. Boring may be by mechanical means, or may be chemical by way of acid secretion which leads to ultimate crumbling of carbonate rocks.

References

1. ANSELL, A. D., 1969: A comparative study of bivalves which bore mainly by mechanical means. American Zoologist, **9**, 857—868.

2. COBB, W. R., 1969: Penetration of $CaCO_3$ substrates by the boring sponge Cliona. American Zoologist, **9**, 783—790.

3. DRZAL, M., and B. SMYK, 1968: The role of the micro-biological element in the formation of structures and forms of a rocky substratum (in polish, english summary). Przeglad Geograficzny, **XL** (2), 425—430.

4. EVANS, J. W., 1968: The role of Penitella penita (CONRAD, 1837) (Family Pholadidae) as eroders along the Pacific Coast of North America. Ecology, **49** (1), 156—159.

5. GILL, W. R., and G. H. BOLT, 1955: Pfeffer's studies of the root growth pressures exerted by plants. Agronomy Journal, **47**, 166—168.

6. GOLUBIč, S., 1969: A comparative study of bivavles which bore mainly by mechanical means. American Zoologist, **9**, 857—868.

7. HENDERSON, M. E. K., and R. B. DUFF, 1963: The release of metallic and silicate ions from minerals, rocks and soils by fungal activity. Jour. of Soil Science, **14** (2), 236—246.

8. KAUFFMANN, J., 1960: Corrosion et protection des pierres calcaires des monuments. Corrosion et Anticorrosion, **8** (3), 87—95.

9. KELLER, W. D., 1957: The principles of chemical weathering. Columbia, Mo.: Lucas Bros. Publ., 111 p.

10. KELLER, W. D., and A. F. FREDERICKSON, 1952: Role of plants and colloidal acids in the mechanism of weathering. Am. Jour. Sci., **250** (594—608).

11. MILLOT, G., et J. COGNÉ et al., 1967: La maladie des grès de la cathédrale de Strasbourg. Bull. Serv. Carte géol. Alsace-Lorraine, **20** (3), 131—157.

12. NEUMANN, A. C., 1966: Observation of coastal erosion in Bermuda and measurements of the boring rate of the sponge Cliona lampa. Limnology and Oceanography, **11** (1), 92—108.

13. POCHON, J., and D. JATON, 1967: The role of microbiological agencies in the deterioration of stone. Chemistry and Industry, Sept. 1967, 1587—1589.

14. SCHAFFER, R. J., 1932: The weathering of natural building stones. Dept., Sci. Ind. Res., Bldg. Res., Special Rept., **18**, 1—149.

15. SCHATZ, A., N. D. CHERONIS, V. SCHATZ, and G. S. TRELAWNY, 1954: Chelation (sequestration) as a biological weathering factor in pedogenesis. Proc. Penn. Acad. Science, **XXVIII**, 44—51.

16. SILVERMAN, M. P., and H. L. EHRLICH, 1964: Microbial formation and degradation of minerals. Advances in Microbiology, **6**, 153—206.

17. SILVERMAN, M. P., and E. F. MUÑOZ, 1970: Fungal attack on rock: solubilization and altered infrared spectra. Science, **169**, 985—987.

18. STOLZY, L. H., and K. P. BARLEY, 1968: Mechanical resistance encountered by root growth habits of plants. Soil Science, **105**, 297—301.

19. TAYLOR, H. M., and E. BURNETT, 1964: Influence of soil strength on the root growth habits of plants. Soil Science, **98**, 174—180.
20. TAYLOR, H. M., and L. F. RATCLIFF, 1969: Root growth pressure of cotton, peas, and peanuts. Agronomy Journal, **61**, 398—402.
21. VOÛTE, C., 1969: Indonesia, Geological and hydrological problems involved in the preservation of the monument of Borobudur. UNESCO Rept. Ser. No. 1241/BMS. RD/CLT; Paris, May 1969, 37 p.
22. WARME, J. E., 1969: Marine borers in calcareous rock of the Pacific Coast. American Zoologist, **9**, 783—790.
23. WEBLEY, D. M., et al., 1963: The microbiology of rocks and weathered stones. Jour. of Soil Science, **14** (1), 102—112.
24. WINKLER, E. M., 1966: Weathering agents for building and monumental stone. Engineering Geology, **5**, 381—400.

9. Natural Rust on Stone

9.1. Iron Content of Minerals

The iron content of the earth's crust averages 5%. At the earth's surface iron is tied up as green or black ferrous-ferric iron in the ferromagnesian silicates, as the black ferrous-ferric oxide magnetite, as the yellowish ferrous sulphides, pyrite and marcasite; as the grey to dark-brown ferrous carbonate siderite, and as the red or black ferric oxide hematite. The last is not only a common pigment in rock but can also accumulate as our most important iron ore. In humid atmospheres the brown to ochre-brown ferric hydroxide, goethite, is the most important mineral of the common "rust", alpha-FeOOH. Goethite is usually accompanied by amorphous (not yet crystallized) ferric hydroxide of the same color. Natural rust is summarized as the "mineral" limonite which is not a mineral in the true sense. All the minerals mentioned here tend to adjust to the humid or semi-humid atmospheric surface conditions as they weather to ferric hydroxide or limonite. Metallic iron also changes to rust, mostly amorphous ferric hydroxide with some magnetic brown maghemite, gamma-Fe_2O_3. Crystallization or aging of the non-crystalline ferric hydroxide leads to the formation of submicroscopic goethite. In some rare instances deep orange gamma-FeOOH, lepidocrocite, can form.

The stone industry is much interested in the stability of the iron minerals and minerals containing iron, especially in their rate of oxidation to ferric oxide and ferric hydroxide which may show as undesirable discoloring of existing colors. The minerals and their behaviour are as follows:

9.2. Ferromagnesian Silicates

Biotite, the black mica, loses its ferrous-ferric iron from the silicate lattice at the beginning of the weathering process because the iron is only loosely built into the lattice. Iron almost immediately precipitates as a natural rust halo nearby the biotite flakes. The presence of larger quantities of biotite flakes evenly distributed across the rock substance may spread an almost even ochre cast over the rock. The release of iron from mica is about 500 times greater in distilled water than from other ferromagnesian silicates leached under identical conditions (KELLER, et al. 1963). The presence of exposed black mica appears to be a source for the rapid release of iron.

Hornblende, augite: These minerals include a variable amount of iron, about 7—15%, as black or green ferrous-ferric iron. Weathering of this mineral group also releases iron, but at a much slower rate than black mica does; it would take several human generations to notice discoloring. The presence of fresh minerals of this group does not appear to threaten discoloring of the surroundings of the mineral grains for at least a century.

9.3. Pyrite, Marcasite

Both iron sulfides, FeS_2, pyrite can occur in almost all types of rocks, whereas marcasite is restricted to sedimentary rocks. Both minerals when exposed to air dissociate and oxidize equally fast to rust, releasing sulfurous or sulfuric acid or both. Surface etching by released sulfuric acid can be frequently observed in marbles and other carbonate rocks. (See photo of marble cross with pyrite-quartz vein Fig. 118 in the chapter on Chemical Weathering.) Greenish tarnish on pyrite is rare under strictly oxidizing conditions, but may show up if an abundance of pyrite is exposed by which the forming rust is reduced again by the pyrite. SINGER and STUMM (1970) summarize the oxidation process of the iron sulfides in respect to the development of acid mine waters which process readily applies to exposed stone surfaces. The basic reactions are as follows:

Stage 1: FeS_2 (solid) plus O_2 goes to Fe^{2+} plus sulfur compound.
Stage 2: Fe^{2+} plus O_2 (aqueous) oxidizes to Fe^{3+}.
Stage 3: Fe^{3+} plus H_2O precipitates as $FeOOH$ (solid rust).
Stage 4: Fe^{3+} plus FeS_2 (solid) reduces to Fe^{2+} plus SO_4^{-2} (green tarnish on pyrite surface).

Stage 1: The solid ferrous sulfide takes up oxygen as soon as the sulfide is in contact with rainwater and the atmosphere, leading to both ionized ferrous and ferric irons plus the sulfate ion. This process continues as oxidation of the ferrous iron takes place in the next stage.

Stage 2: Oxidation of the ferrous iron (Fe^{2+}) takes place from oxygen present in rainwater. But before such oxidation occurs, migration of the ionized Fe^{2+} may occur to a short distance from its source where it then becomes oxidized to the more stable but still-ionized Fe^{3+}.

Stage 3: Fe^{3+} precipitates, first as amorphous ferric hydroxide (rust). This slowly crystallizes to the microcrystalline ochre-brown mineral goethite, rarely to the deep red-orange mineral lepidocrocite. The solubility of ferric hydroxide in water is extremely small, only about 0.0005 ppm at pH = 5, and still less at pH = 7 (0.000005 ppm). Fig. 134 presents the solubility of ferric hydroxide against the pH of the solvent. The release of sulfuric acid during weathering of pyrite may temporarily lower the pH of the attacking rainwater to near pH = 3, at which acidity the solubility of ferric hydroxide increases rapidly to about 5 ppm, and 550 ppm at a pH of 2; this acidity, however, is not believed to be reached in nature on a stone surface. The rapid dissipation of the iron away from the dissolving pyrite does generally not permit the contact of the precipitated ferric hydroxide with the still unoxidized portions of the pyrite grains. The observed halos up to 3 to 4 inches away from the source material suggest that some iron had to travel despite the practical insolubility of the ferric hydroxide at average rainwater acidities. The presence of iron and sulfur bacteria is believed to accelerate the oxidation process by a factor of up to 10^6.

Stage 4: If the pyrite masses are large enough so that the precipitated ferric hydroxide is in contact with the original ferrous sulfide, the strongly reducing

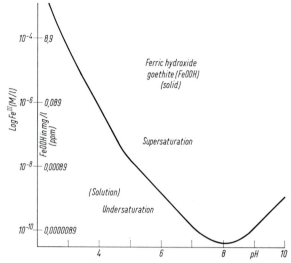

Fig. 134. Solubility curve of ferric-hydroxide, natural rust. After STUMM and LEE (1960)

effect of pyrite may reduce the Fe^{3+} back to Fe^{2+} where it forms greenish ferrous sulfate — in spite of strongly oxidizing environments. Oxidation of ferrous sulfide grains in nature may start after only a few years of surface exposure.

9.4. Siderite

Ferrous carbonate occurs occasionally as grey or light brown grains in some marbles and other carbonate rocks. Siderite is much less soluble than calcite or dolomite; the solubility depends on the CO_2 partial pressure of the atmosphere or of the attacking waters. Siderite oxidizes at the surface upon contact with rainwater to rust-like spots which resemble those after the weathering of pyrite. The yellowing of many crystalline white marbles by so-called "aging" may originate from siderite. Oxidation of small grains of siderite in the Pentelian marble on the Parthenon of Athens is cited by KIESLINGER (1932).

Some small siderite nodules enclosed in exposed architectural aggregate have developed halos of ferric hydroxide within two years of exposure on exterior floor slabs on the mall of the Notre Dame Memorial Library. Their solution and subsequent oxidation causes a spread of 3 to 4 inches to the north-east as the main direction of the blowing rain hits from the southwest. The process of solution-transport-oxidation has been discussed above in connection with iron sulfides.

9.5. Hematite

Hematite (alpha-Fe_2O_3) and the magnetic maghemite (gamma-Fe_2O_3) are very stable minerals. Hematite usually forms during oxidation at the higher temperatures of warm climates; hydration does not occur readily if hematite is well crystallized. Hematite is also a component of common rust.

9.6. Industrial Rust

The composition of rust derived from metallic iron or steel alloys depends much on the kind of crystallization of the original material. All industrial ferro-metals except certain stainless steels are subject to the rusting process, even in deserts. The minerals of industrial rust are generally microcrystalline hematite, the non-magnetic alpha-Fe_2O_3; goethite, nonmagnetic alpha-$FeOOH$; maghemite, magnetic gamma-Fe_2O_3, and (mostly) amorphous ferric hydroxide. Many industrial rusts are thus magnetic.

9.7. Physical Damage by Rust

Stone anchors and bolts as well as rods in reinforced concrete have been much exposed to damage by rust burst. The same rust burst may be observed during the oxidation of pyrite. The calculation of the volumetric expansion from metallic

Table 34. *Volume Expansion of Iron and Iron Minerals to Rust*

Original Material Source of Rust	D (g./cm.³)	Volume Exp. to Hematite (%) D = 5.26	Volume Exp. to Goethite D = 4.37	Estimated True Expansion
Structural steel	7.5	30%	42%	50—60%
Pyrite	5.02	—4.8	13%	20—30%
Marcasite	4.89	—7%	11%	15—30%

iron to rust is very difficult because most of the freshly formed rust is still amorphous $FeOOH$. The attempt is made in Table 34 to calculate the theoretical volume expansion of metallic iron, pyrite, marcasite to both the crystalline goethite and to common rust. The estimated true expansion is a correction of expansion values for unknown quantities of amorphous ferric-hydroxide in rust.

Summary

The weathering of iron minerals and of minerals which have iron built into the crystal lattice, like the ferro-magnesian silicates, releases iron to the immediate surroundings of the mineral grain. The process of oxidation and hydration forms nearly insoluble oxides and hydroxides of iron as rusty stains. Limonite, a collective mineral term for all iron hydroxides both crystalline and amorphous, is the stable weathering end product in humid and semi-humid climates.

In silicates, biotite loses iron first, whereas the hornblende is relatively stable. The water soluble ferrous carbonate dissolves quite readily in water high in CO_2 in polluted urban atmospheres; if not transported the iron will soon precipitate as ferric-hydroxide nearby, first amorphous aging later on to finely crystalline FeOOH, the goethite.

References

1. KELLER, W. D., W. D. BALGOARD, and A. L. REESMAN, 1963: Dissolved products of pulverized minerals, Part I. J. Sed. Petrology, **33** (1), 191—204.

2. KIESLINGER, A., 1932: Zerstörungen an Steinbauten, ihre Ursachen und ihre Abwehr. Leipzig-Wien: Franz Deuticke.

3. SINGER, P. C., and W. STUMM, 1970: Acidic mine drainage: The rate-determining step. Science, **167** (Febr. 20), 1121—1123.

4. STUMM, W., and G. F. LEE, 1960: The chemistry of aqueous iron. Schweiz. Zeitschr. Hydrologie, **22**, 295.

10. Fire Resistance of Minerals and Rocks

Fire damage has frequently created problems with building stone and concrete. Urban fires have severely damaged stone, especially granites and quartz-sandstones, but also limestones, dolostones and marbles. Major conflagrations in many European cities during World War II have left behind numerous ruins as cases for detailed studies.

10.1. Minerals

Cracking has been attributed to the difference of thermal expansivity of minerals. Uneven volumetric or linear expansion of minerals may cause disruption of stone or concrete in fire. All minerals expand with increasing temperature.

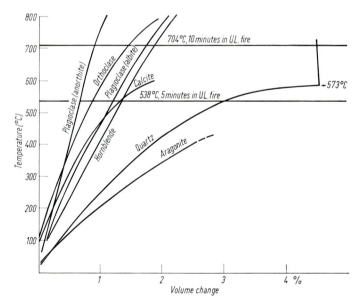

Fig. 135. Volume expansion of some common rock forming minerals. The U. L. Standard Curve is plotted into the graph for 5 minutes of heating at 538° C, and 10 minutes at 705° C. Data plotted from SKINNER (1966)

The volume change of a few common rock-forming minerals is presented in Fig. 135. The temperature scale along the ordinate axis of the graph marks three important points, i.e., 538° C, the temperature reached in a "Standard Fire" in an enclosed

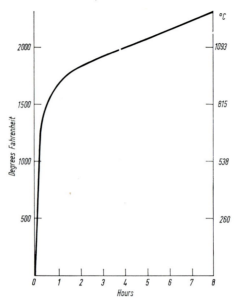

Fig. 136. Standard heat curve of Underwriter's Laboratory in a closed room with thermocouples implanted in walls and ceilings. The temperatures are obtained from readings of not
less than nine thermocouples for a floor, roof or partition, and not less than eight thermocouples for a structural column symmetrically distributed to show the temperature near all
parts of the sample. From UL ⧻ 263 (1959)

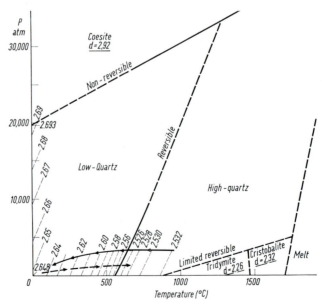

Fig. 137. P-T-V diagram of silica. Dashed lines with arrows mark process during a fire; full
line with arrows marks probable course of cooling from liquid magma to solid rock. Modified
from Boyd and English (1960)

room after 5 minutes (Underwriters Laboratory, 1964), 704° C, the temperature reached after 10 minutes (Fig. 136); 573° C, the temperature of the phase change from low- to high-quartz.

Quartz expands about four times more than the feldspars, twice as much as hornblende. Quartz is therefore considered as the most critical mineral under conditions of intense heating. KIESLINGER (1954) ascribes the fire damage to the spontaneous conversion of low- to high-quartz. BOYD and ENGLISH's (1960) updated P-T-V diagram of Fig. 137 indicates the degree of volumetric expansion in terms of decrease of density during the process of heating. Starting with the zero point, the densities decrease with increasing temperatures but increase with increasing pressure. Theoretical calculations indicate that the lines of equal densities run about parallel to the phase boundary between low- and high-quartz. The density decrease accelerates almost exponentially towards the phase boundary, with rising temperatures, to a total of about 3.7% from room temperature to the boundary line, but only 1% across the phase boundary from low- to high-quartz. The phase conversion is facilitated by the very similar crystallography of both the low- and the high-quartz. Further heating and recrystallization from high-quartz to tridymite and cristobalite is a slow geological process and should be thus considered as "limited reversible". Coesite is the form of silica recrystallized by shock, such as by volcanic explosion or meteorite impact. Phase boundaries marked with a full line are based on laboratory test results. The behaviour of granite in fire as well as the origin of micro-crack porosity may be explained with the help of the P-V-T diagram of quartz, as follows: When quartz expands during heating it starts exerting pressure against its surroundings (follows dashed line in Fig. 137). The volume increase to about 500° C produces pressures of about

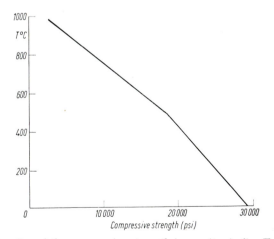

Fig. 138. Deterioration of the compressive strength in granites in fire. From TARR (1915)

1000—2500 atm. The phase boundary moves upward from 573 to near 600° C under stress. Rocks with unfavorable tensile strength may crack well below 500° C; rock disintegration is then readily completed with another per cent volume increase across the phase boundary. Uneven quenching with cold water from fire hoses can occasion surface spalling by creating great internal differential stresses. Spalling and cracking can occur well below 573° C.

TARR (1915) stressed preheated Missouri granites and found linear decrease of the compressive strength from an original 29,000 psi at room temperature to 18,000 at about 573° C, with more loss of strength above 573° C, according to Fig. 138. Quartz-sandstones crack in fire in a similar way as granites. Fig. 139 pictures an example for spalling of a sandstone pillar in fire.

Fig. 139. Fire damage by surface spalling on pillar of quartz-sandstone blocks. Building next to Mainz Cathedral, Germany. Photo taken 1953

Calcite shows linear thermal expansion parallel to the c-axis, about 2%, but contracts about 0.5% perpendicular to the c-axis. The plot of the linear expansion of some common rock forming minerals shows this (Figs. 140 and 141). Simultaneous expansion and contraction of calcite in crystallographically oriented marbles may be disruptive even at temperatures below 100° C. ROSENHOLTZ and SMITH (1949) found that heating of marble causes expansion, more so during the first heating cycle than during later cycles (Fig. 142). The high degree of expansion at temperatures below 300° C during the first heating cycle is explained as partial relief of residual stress of prestressed marble. This phenomenon should be expected with all crystalline marbles which were prestressed during metamorphism sometimes in the geologic past. Prestressed metamorphic non-carbonate rocks failed to expand visibly during the process of milling.

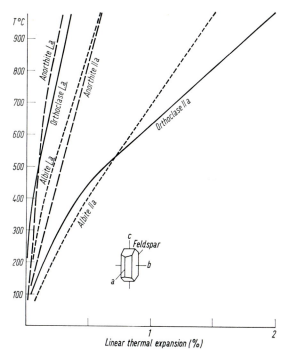

Fig. 140. Linear expansion of the feldspar group

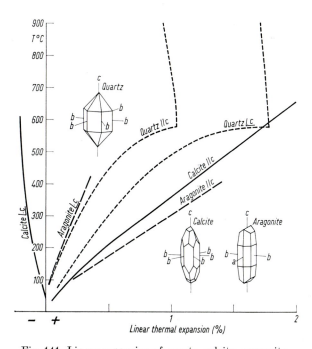

Fig. 141. Linear expansion of quartz, calcite, aragonite

Superficial calcination of carbonate rocks should be expected in only very intense fires as calcite begins to dissociate near 900° C, dolomites at near 800° C. A thin layer of burnt lime insulates the undamaged stone underneath, but leaves shallow scars after the soluble calcium hydroxide has washed off.

Table 35. *Thermal Expansion of Quartz from 0° C*

Temperature (° C)	Change of Length, %		Change of Volume, %
	⊥ to c	‖ to c	
50	.07	.03	.17
100	.14	.08	.36
200	.30	.18	.78
300	.49	.29	1.27
400	.72	.43	1.87
500	1.04	.62	2.70
570	1.46	.84	3.76
573	transition from low- to high-quartz		
580	1.76	1.03	4.55
600	1.76	1.02	4.54
700	1.75	1.01	4.51
800	1.73	.97	4.43
900	1.71	.92	4.34
1000	1.69	.88	4.26

From: SKINNER (1966).

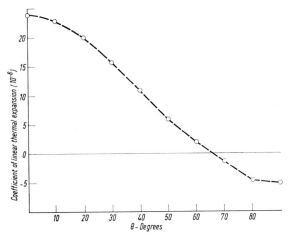

Fig. 142. Thermal expansion coefficient of Yule marble, in two heating cycles. After ROSEN-HOLTZ and SMITH (1949)

10.2. Rocks

Rocks which are low in carbonate minerals and quartz and also low in orthoclase and Na-plagioclase, expand only very little (GRIFFITH, 1936). Basic igneous
rocks should be preferred where exposure to high temperatures is expected. Thermal
expansion values of common rocks are summarized in Table 12 of the chapter Stone
Properties.

All rocks should be expected to spall along the boundary of heated to unheated
rock substance as the stresses become a maximum along this line where the thermal gradient is greatest. Ornaments, ridges and other projections along the stone
surface are especially vulnerable, much more so than flat surfaces.

Summary

Exposure of rock to intense heating results in differential expansion of the
minerals. Acid igneous rocks and sandstone rich in quartz, orthoclase and sodium
plagioclase are more susceptible to fire damage than rocks without these minerals.

Quartz is the most expansive mineral with 3.76% volume expansion from
room temperature to 570° C, but only 0.7% across the phase boundary from
low-quartz to high-quartz. Disruption of rocks may occur at the boundary of
heated to unheated rock, especially in quartz-sandstones and quartz-containing
granitic rocks. Anomalous contraction of the quartz during the cooling process
from the liquid magma to the cooled rock in the geologic past resulted in micro-
cracking of granitic rocks leading to high micro-porosity and relatively low strength,
lower than the strength of other igneous rocks.

Calcite expands along the c-axis while it contracts perpendicular to it, stressing
marbles which have oriented calcite grains. Very intense heating of limestones and
marbles to near 900° C, dolomites to 800° C may cause superficial burning which
is soon followed by slaking, leaving ugly shallow scars.

References

1. BOYD, F. R., and J. L. ENGLISH, 1960: The quartz-coesite transition. Journal Geophysical
   Research, **65** (2), 749—756.
2. KIESLINGER, A., 1954: Brandeinwirkungen auf Naturstein. Schweizer Archiv, **20**, 305—308.
3. ROSENHOLTZ, J. L., and D. T. SMITH, 1949: Linear thermal expansion of calcite, var.
   Iceland spar, and Yule marble. The American Mineralogist, **34**, 846—854.
4. SKINNER, B. J., 1966: Thermal expansion. In: Handbook of Physical Constants (S. P.
   CLARK, ed.), Geol. Soc. America Memoir **97**, 75—96.
5. TARR, W. A., 1915: A study of some heating tests, and the light they throw on the cause of
   the disintegration of granite. Economic Geology, **10**, 348—367.
6. Underwriter's Laboratory, 1964: Fire tests of building construction and materials, Standard of Safety. Underwriter's Laboratory, Publication # 263 (1959, revised 1964), 22 p.

11. Frost Action on Stone

Frost damage to stone and concrete in moderate humid and semi-humid climates has long been known as a major disruptive agent which deserves attention. Frost action results from a combination of factors, such as spontaneous volumetric expansion from the water phase to the ice phase, the degree of water saturation of the pore system, the critical pore size and the continuity of the pore system.

11.1. The Process of Freezing

The pressure-temperature phase diagram of water and the various phases of ice has been known for over a hundred years. The phases water, ice I and ice III may occur in nature up to and near the "triple point". Fig. 143 enlarges that portion of

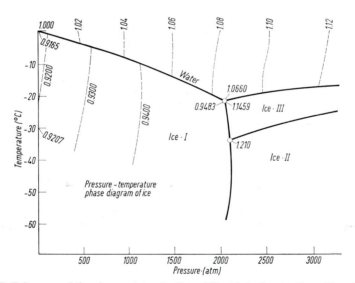

Fig. 143. P-T diagram of the phases water, ice I, ice III with isodensity lines. From WINKLER (1968)

the ice diagram which occurs in nature, with contour lines, connecting points of equal density for both the water and the ice I phase (WINKLER, 1968). This updated diagram offers basic theoretical information on frost danger which should be interpreted as follows: At 0° C water freezes at 1 atmosphere pressure. The pressure increases about 1750 psi for each degree centigrade of temperature decrease during confined freezing. According to the ice-water equilibrium curve, ice can reach a maximum pressure of 610 atm at —5° C without melting, 1130 atm at —10° C, and 2115 atm at —22° C, the "triple point" at which ice of $d = 0.948$ reverts either to denser water of $d = 1.086$, or to much denser ice III with $d = 1.1459$. The conversion of the less dense ice I to the always denser water thus prevents the development of higher pressures than the curve indicates. At further temperature decrease, the stress increase is only small, with the maximum occurring at —40° C and 2,120 atm; further temperature decrease results in a slight recovery of stresses. Ice skaters notice a rapid increase of the ice hardness with decreasing temperature; this ice property, however, is of little interest to the stone industry.

11.2. Frost Danger to Rocks

The disruptive force of ice can be explained with the help of the P-V-T diagram of ice; the different sensitivity of various rocks to frost damage cannot find its explanation so simply. KIESLINGER (1930) summarized and evaluated ice action and its geological importance. Today the porosity, critical saturation, critical mean pore size, continuity of pores, and others are recognized as the determining factors for the susceptibility of rock to freezing. The complexity is enhanced in concrete where cement and cement interfaces with the aggregate stone are to be considered. The complexity and wealth of information has also brought forth different ideas on frost damage to stone and concrete as follows:

a) *Hydraulic pressure:* POWERS (1955) ascribes the disruptive force of ice not merely to the direct expansive action from water to ice but rather to the hydraulic pressure generated during the initial freezing, whereby still unfrozen water is displaced away from the advancing front of frost.

Fig. 144. Protection of quarry face from winter frost with straw retained by snow fences. Mankato dolomite, Mankato, Minnesota

b) *Temperature-dependent volume change of adsorbed water:* DUNN and HUDEC (1965) have found that not all the water in pore systems freezes at or near the freezing point, but that some remains liquid at temperatures as low as —40° C. Evidently, ordered water is held so tightly that freezing is difficult. The presence of clay minerals distributed in carbonate rocks enables capillaries to attract water tightly and probably in an ordered fashion; the clay crystals act as the negative pole attracting the positive end of the water molecule. This tight bond may develop pressures of still unknown magnitude. DUNN and HUDEC (1966) found dolomites with clay as impurity to be more sensitive to frost action than limestones with an equivalent quantity of clay (Fig. 144). Rejection of the clay particles during the process of recrystallization of dolomites tend to form continuous bands between the tiny dolomite grains as "rejection texture" which clay bands tend to transport and adsorb water. Rejection textures were not observed with limestones. Tightly adsorbed water in an ordered fashion is believed to disrupt the rock rather than the direct action of freezing. DUNN and HUDEC (1966) also show that unfrozen water well below freezing temperatures expand enough to aid in the disruption process. Fig. 89 of chapter 6 Moisture and Salts in Stone indicates support of this theory. Frost damage may be thus inversely proportional to the amount of freezable water.

c) *Dual mechanism theory* by the adsorption of unfrozen bulk water: LARSEN and CADY (1969) have proven with laboratory experiments that the destructive effect of frost action is the result of hydraulic pressure generated by an increase in the specific volume of water during the change of state from bulk water to adsorbed water in addition to direct action of ice. Coarse-grained and coarse-pored rocks generally withstand freezing well, though they may not comply with the desired standards. Fine-grained rocks which have sorbed water in vacuum are generally very susceptible to frost above 5%, but are very durable below 1% water sorption. WALKER *et al.* (1969) determined the critical pore size for the freezing-thawing durability to be about 5 microns. Larger mean pore diameters permit outward drainage and escape of fluid from the frontal advance of the frost line.

11.3. Influence of Water Saturation and the Number of Freeze Cycles

LARSEN and CADY (1969) state that specimens which were soaked continuously before freezing were more susceptible than specimens which were soaked and subsequently dried at 75° F and 50% relative humidity. This experience appears

Fig. 145. Block of Indiana limestone with frost damage in upper layers. Spalling along bedding planes near top of block to where moisture has travelled by capillary action before stone could cure

to run parallel with the observed greater frost sensitivity of quarry-moist stone blocks as compared with cured stone (Fig. 145). The Romans were already well aware of this stone property. The curing of English Portland stone was carefully practiced by Sir Christopher Wren, London's famous Baroque architect. In rip-rap, bridges and seawalls, exposure to frost after continuous soaking with water is rare but should be expected to occur just above the water level, as exemplified in Fig. 146.

Fig. 146. Frost damage on pier of Lake Michigan. The lake level is located about 3 feet below the surface. Damage is a maximum along the outer edge to which most of the moisture was drawn. Rear block was built up to original surface by recent repair. Michigan City, Indiana

Fig. 147. Dolomite stone in low wall disintegrates along bedding planes, probably along films of clay layers. Frost action, combined with salt action or ordered water may be responsible for the damage. The wall was erected about 100 years ago. Wall in Milwaukee, Wisconsin

POWERS (1955) pointed out that a single slow-freeze cycle can give good infor-mation on the durability of stone and concrete as an alternative to the 300 freeze-thaw cycles recommended by the ASTM Standards. LARSEN and CADY (1969) carried out systematic tests to develop a one-cycle slow-freeze test by correlating results with the durability factor obtained from the existing ASTM Method C-290 (the standard freezing and thawing test in water at 100 freeze-thaw cycles average). DUNN and HUDEC (1966) claim acceleration of the freeze-thaw process in a 10% NaCl solution with a factor of more than 10 over the severity of freezing and thawing of pure water. Fig. 147 may be an example of this effect in practice.

11.4. Durability Factor

The ASTM C-290, 291-67 Standard Method of Test for Resistance of concrete specimens to rapid freezing in air and thawing in water for 300 cycles or until the relative modulus of elasticity has decreased to 60% of the original modulus. The freezing-thawing cycle for this method consists of alternately lowering the temperature from 4.4 to $-17.8°$ C in a period not more than 3 hours and from -17.8 to $4.4°$ C in a period not more than 1 hour. The durability factor can be calculated from

$$DF = \frac{P N}{M} \text{ where}$$

$DF =$ durability factor of the test specimen,
P = relative dynamic modulus of elasticity after N cycles, in per cent,
N = number of cycles at which P reaches the specific minimum value for discontinuing the test, and
M = specific number of cycles at which the exposure is to be terminated.

The method of test is also applicable to stone. WALKER's (1969) one-cycle freezing test appears to be applicable as a screening test, while the durability factor can be considered as a simple, rapid and inexpensive method of potential field performance.

Summary

Maximum theoretical ice pressures are presented and the P-V-T diagram of ice discussed, as ice is believed to be a major factor in the disruption of stone and concrete. Ice pressures rapidly increase with decreasing temperature to about $-22°$ C where ice can exert a maximum of 30,000 psi pressure. Rocks with large pore diameters are less frost sensitive than rocks with very small mean pore size. A critical mean pore size of 5 microns renders stone frost sensitive. Stone disruption may be explained by the volume increase of water to ice, by the displacement of pore water away from the advancing frost, by the conversion of pore water to ordered water, and by the volume increase by unfrozen water at temperatures below freezing.

References

1. DORSEY, E. N., 1940: Properties of ordinary water substance. New York: Reinhold Publishers, 673 p.
2. DUNN, J. R., and P. P. HUDEC, 1965: The influence of clay on water and ice in rock pores. Physical Research Report RR 65-5, New York State Dept. of Public Works (1965).
3. DUNN, J. R., and P. P. HUDEC, 1966: Water, clay and rock soundness. Ohio Journal of Science, **66** (2), 153−167.

4. KIESLINGER, A., 1930: Das Volumen des Eisens. Geologie und Bauwesen, **2**, 199—207.

5. LARSEN, T. D., and P. D. CADY, 1969: Identification of frost-susceptible particles in concrete aggregate. National Cooperative Highway Research Program Report, **66**, 62 p.

6. POWERS, T. C., 1955: Basic considerations pertaining to freezing-and-thawing tests. American Society for Testing Materials, Proceedings, **55**, 1132—1155.

7. WALKER, R. D., H. J. PENCE, W. H. HAZLETT, and W. J. ONG, 1969: One-cycle slow-freeze test for evaluation aggregate performance in frozen concrete. National Cooperative Highway Research Program Report, **65**, 21 p.

8. WINKLER, E. M., 1968: Frost damage to stone and concrete: Geological considerations. Engineering Geology, **2** (**5**), 315—323.

12. Silicosis

12.1. Dust

Dust from mining and quarrying operations has plagued man's respiratory system for thousands of years. Exposure to quartz dust has caused most of the trouble, because it may lead to incurable or even fatal silicosis. The American Public Health Service, Committee on Pneumoconiosis, defines silicosis as "A dust desease due to breathing air containing silica (crystalline SiO_2), characterized anatomically by generalized fibrotic changes and the development of miliary nodulation in both lungs, and clinically by shortness of breath, decreased chest expansion during breathing, lessened capacity for work, absence of fever, increased susceptibility to tuberculosis (some or all of which symptoms may be present), and by characteristic X-Ray findings (FORBES, 1950)".

Some other dust diseases of the lungs are known to exist besides silicosis, such as anthracosis by anthracite coal dust, asbestiosis by asbestos dust, and the very dangerous berylliosis by beryllium or beryllium oxide dust. All these diseases are collectively called Pneumoconiosis. Only silicosis is discussed in the following;

Table 36. *Number of Diseased Workers, in per cent, for Different Stages of Silicosis, in the Lead-Zinc District of Missouri, Oklahoma, Kansas*

Stage of Disease	No. of Cases %	Ave. Duration of employment
First stage silicosis	16.6	5.3 years
Second stage silicosis	19.7	6.7 years
Third stage silicosis	9.4	8.0 years
Silicosis combined with tuberculosis	14.3	8.0 years

From: Silicosis and Allied Disorders, Anonymous (1937).

other varieties of pneumoconiosis do not generally occur in the commercial stone industry and are here omitted. The subject is well reviewed by FORBES (1950), also by DAVIL et al. (1935). The following factors may contribute to the disease according to standards of the American Public Health Service:

1. The silica dust particles finer than 5 microns are able to penetrate the air passages of the lungs. The particles retained in the lungs generally range from 1 to 5 microns in size.

2. The maximum permissible concentration of quartz in the dust is 5 million particles per cubic foot of air.

3. Silicosis stems from the slow accumulation of dust over a period of usually 12 to 18 years; it may develop faster in case of unusually high dust concentration or high individual sensitivity. Silica concentrations of 3% or higher may be effective.

Statistical studies on silicosis were performed with 720 miners in the Tri-State Lead and Zinc District of Oklahoma, Missouri and Kansas in 1913. Of these, 45.7% had contacted silicosis in rocks with 90 to 97% silica content, mostly chert. The break-down of the figures is given in Table 36.

No recent cases have been reported in the Tri-State district, where underground operations are now well ventilated, and drilling and crushing operations are continuously sprayed with water.

12.2. Dust-Producing Materials

The U.S. Public Health Service has conducted a survey of quartz content in industrial dust-producing raw materials (Table 37).

The composition of granite-cutting dust varies with the quartz content from 5 to 35%. Quartz-sandstones, quartzite, and chert rock produce pure silica dust. Rocks composed of silicate minerals do not present any danger as the silica is

Table 37. *Dust-Producing Materials and Their Quartz Content*

Dust Producing Material	Quartz, %
Bituminous coal mines	54.0 (from sandstone in wallrock)
Anthracite mines	31.0 (from sandstone in wallrock)
Granite quarries (granite high in quartz)	35.2
Granite quarries (granite low in quartz)	5.0 or more
Cement plant	6.5
Slate mill (Vermont red slate)	3.0
Slate mill (Vermont green slate)	trace
Marble quarries and mills	none

From: Anonymous (1937), GARDNER (1938).

tightly built into the crystal lattice with other elements. Limestone and marble dusts disappear from the lungs without any visible change; these dusts rather help localize portions of the lungs which are inflamed by tuberculosis. Dusts of anthracite, bituminous coal, hematite, precipitator ash and soap stone develop inert deposits in the lungs which remain fixed in the tissues.

12.3. Dust Travel in the Lungs

Dust enters the lungs through the bronchial tree, the Y-junction of the wind pipe leading to the lungs. Free dust, 0.5 to 5 microns in diameter, is consumed by phagocytes in the alveoli, dust-consuming cells in the air cells of the lungs. The nature of the dust determines the characteristics of the phagocytes which, in turn, influence the symptoms of the disease. Dust rapidly scatters throughout the air passages and along the walls of the alveoli within minutes after entry, and there it becomes ingested by phagocytes immediately within the alveoli (HEPPLESTON, 1954). The phagocytes come to rest in the interstices of the alveoli which are connected directly to the bronchioles (minute thin-walled branches of the bronchia) where phagocytes accumulate to foci (centers) and develop further to macrophages, major concentrations of dust-consuming cells. Some of the phagocytes and residual free dust may move back up the bronchial tree by the mechanism of "alveolar clearance" to become expectorated and so removed from the body. KLOSTERKOETTER (1957) demonstrates with rats that physical exercise accelerates alveolar clearance. Scar tissue stiffened by existing foci (centers) diminishes the efficiency of alveolar clearance: persons with a history of exposure to dust are more susceptible to pneumoconiosis (dust lung) than are persons with no such disease in their medical histories. The further development of semi-elastic fibres (fibrosis) surrounding the foci or centers lead to emphysema: the concentration and expansion of the originally smooth air passages during breathing are shortened and narrowed. Emphysema is characterized by shortness of breath and gasping for air.

12.4. The Physiology of Silicosis

A dissected silicotic lung appears to be large and rigid, with pigmentation of grey to black depending on the amount of other dusts absorbed along with silica. In earlier stages the lung is smooth and glistening; in later stages, however,

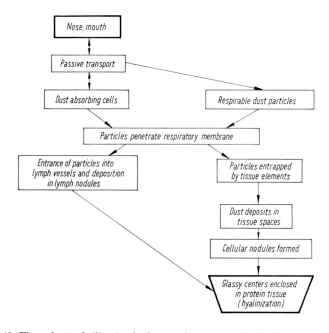

Fig. 148. Flow chart of silica in the human lungs, simplified after LANZA (1963)

adhesions of hard nodules 2 to 4 mm. in diameter are frequent. Their distribution throughout the lungs may be even, or may be concentrated in the upper portion of the lungs. The lymph nodes are enlarged, black, and hard, adhering to the bronchii and vessels and narrowing the bronchia, known, in clinical terms, as bronchial stenosis. The nodules are of concentric arrangement, with reactive enclosing protein tissue around a glassy or hyaline (opal-like) center. The protein zone varies in shape and width, and may be pigmented with other dusts. The total lack of the reactive zone about a silicotic nodule speaks of a quiescent condition. Its presence, however, indicates the progression of acute silicotic processes. The mechanism of dust mobilization is a complex biochemical process which will not be discussed here. Silica dust may free itself again from the nodules towards the free air spaces in the lung, by alveolar clearance, with the help of the cilia, vibrating hairs; a smaller fraction left behind may be carried to lymphatic channels and eventually to lymph nodules. New inflammations will occur by dust remaining in the alveoles, or air spaces. Severe silicotic involvement results in thickening of air passage walls, fibrosis, etc., as was discussed with dust travel in the lungs. Fig. 148 illustrates a simplified sketch of the migration and emplacement of silica in the lung.

About 75% of all advanced cases of silicosis are associated with tuberculosis, according to GROSS (1959) and LANZA (1963). Until recently, the mechanical slicing action of the sharp-edged quartz dust was believed to play the major role in the development of silicosis. Partial laceration of the lung was believed to facilitate the easy entry of the tubercles. Modern research performed by the Air Hygiene Foundation of America, Inc., failed to find any evidence of mechanical lung laceration by quartz dust.

12.5. Signs and Symptoms of Silicosis

The following symptoms are characteristic for silicosis and develop by prolonged exposure to silica dust:

Increasingly labored breathing with decreasing ability to perform physical work is observed as the expansion of the chest diminishes and chest pain develops (tight chest); all available chest muscles have to be used for respiration. Tubercle infection may occur at any time during silicosis; the possibility of infection increases with more advanced cases. It is still unknown why tuberculosis enters the picture and why sputum does not show tubercle bacilli until just before death. In recent years tubercle infection accompanying silicosis has decreased with the general decline of tuberculosis mortality.

Most recent research on silicosis has revealed that only people with previous lung infections are likely to become victims of silicosis (ANONYMOUS, 1965). Only 9 out of 594, or 1.5%, of the workers in the mines of Boliden, Sweden, came down with silicosis without previous lung infections. In contrast, 55%, or 39 out of 71, or 55%, of the workers with previous lung infections had contracted silicosis. The self-clearing mechanism of the lungs generally takes care of some of the dust removal from the lungs.

Silicosis remains stationary as soon as the exposure to silica dust ceases. Lung cancer and pneumonia have not been observed as a consequence of dust exposure.

12.6. Prevention of Silicosis

The elimination of the disease can only be solved with complete dust removal and careful dust prevention in quarries and mines, especially where quartz is known to be present in the rock. Water sprinkling and spraying systems should be installed and frequently inspected. Based on experience in Sweden, new employees should be carefully screened for previous lung troubles. Hope is expressed that the full understanding of the disease, its detection and prevention will soon make silicosis a thing of the past.

Summary

Sandstone, granite, and many other rocks may contain free quartz. The production of dust having in excess of 3% free SiO_2 may affect the lungs and may develop silicosis upon prolonged exposure. Silicotic lungs lack elasticity and strongly reduce breathing capacity and working potential. About 75% of all advanced cases have also contracted tuberculosis. Dust of limestone, marble, and gypsum are considered relatively harmless. Proper dust prevention can reduce the flying dust to a level below the danger limit. Workers with previous lung troubles are likely to contract silicosis much more easily than people with healthy lungs.

References

1. Anonymous, 1965: Silicosis research. Engineering and Mining Journal, **166** (**4**), 98—99.
2. Anonymous, 1937: Silicosis and allied disorders; historical and industrial importance: The Medical Committee. Air Hygiene of America, Inc., 178 p.
3. Davil, G. C., E. M. Salmonson, and J. L. Earlywine, 1935: The Pneumoconioses of the Lungs (Silicosis); 1. Bibliography and Laws, Industrial Medicine, 2. Literature and Laws of 1934. Chicago Medical Press, 493 p.
4. Forbes, J. J., S. J. Davenport, and G. G. Morgis, 1950: Review of literature on dusts. U.S. Dept. of the Interior, Bureau of Mines, Bull. 478, 333 p.
5. Gardner, L. M., 1938: Reaction of the living body to different types of mineral dusts. Am. Inst. of Mining and Metallurgical Engineers, Techn. Publ. No. 929, pp. 1—15.
6. Gross, P., 1959: Silicosis: A critique of the present concept and a proposal for modification. Am. Medical Assoc. Archives of Industrial Health, **19**, 426—430.
7. Heppleston, A. G., 1954: Pathogenesis of simple pneumokoniosis in coal workers. Journal of Pathology and Bacteriology, **67**, 51—63.
8. Klosterkoetter, W., 1957: Tierexperimentelle Untersuchungen über das Reinigungsvermögen der Lunge. Archiv der Hygiene, **141**, 258—274.
9. Lanza, A. J., 1963: The pneumoconioses. New York and London: Grune and Stratton, 154 p.

13. Stone Conservation on Buildings and Monuments

The rapid, visible decay of stone in urban areas has made it necessary to protect stone surfaces from premature decay; the attempt is to halt the natural process of stone decay but we can merely slow down nature's continuous gnawing. Protective commercial chemicals have been applied since the 17th century. The heterogeneity of stone, however, does not permit a generalized panacea; the number of failures in stone preservation is greater than cures. In the early days of stone treatment, the access of surface water was prevented by the application of linseed oil and hot waxes. Total sealing rarely solved the problem, as the actual travel routes for moisture within the monument or from the ground were not yet understood. Undesirable stains, efflorescence and accelerated flaking were the result. Today, the great variety of chemicals consists of sealing paints and varnishes, and silicones, resins and other compounds designed to achieve efficient penetration.

Stone preservatives fall into two general categories, i.e., sealers and hardeners. Many modern compounds serve both functions.

13.1. Sealers

Sealers develop a tight, impervious skin which prohibits the access of moisture. Liquid hot wax and paraffin have been applied for centuries. Moisture acts from behind such insulation, however. Conservators were not aware of the very complex internal travel routes of moisture and insulation against rising ground moisture factors which were still unknown in the 19th century. The action of moisture from behind the insulating surface barrier has accomplished more damage than protection. London's Cleopatra Needle was treated with a mixture of Damar resin and wax dissolved in clear petroleum spirit in 1879 (BURGESS and SCHAFFER, 1952). The treatment of the granite monument prevented the access of moisture to the salts, still entrapped from the Egyptian desert — and has protected the monument. The sister monument in Central Park, New York, has fared less favorably than its London counterpart; the surface treatment was not started until 10 years after its erection in New York, after high relative humidity had already penetrated the stone.

Marble: Treatment of marble surfaces with linseed oil has often resulted in unpleasant yellowing.

Sandstone: Surface seals on sandstone have never been beneficial preventing drying of the stone. Moisture and salts trapped just beneath the stone surface break the skin whereby the temporarily indurated film curls to concave peels.

Carbonates: Carbonate rocks are subject to strong solution attack by rain and moisture. The interaction of the sulfates in the air with the carbonates can form soluble gypsum at the stone surface. If, however, Ca-sulfite crusts can be formed, the stone surface attack is retarded. LEWIN (1968) exposed a variety of limestones to $CaCl_2$, NaF, and other compounds. He found that NaF reacted with the thermodynamically less stable interstitial calcite particles, downward to the depth of maximum fluid penetration; metasomatic replacement took place to form the water-insoluble and harder CaF_2, the mineral fluorite. Such conversion of tiny intergranular calcite particles reduces microchannels and diminishes the absorption. The new growth and intergrowth forms a tight network which is stronger and more cohesive than the parent calcitic substance.

Replacement of calcite with barium hydroxide, in the presence of urea as a catalyst, converts calcite to insoluble barium carbonate, the mineral witherite. Contact with sulfates converts the barium carbonate to the still less soluble barium sulfate, barite.

The treatment of limestone and marble with NaF and $Ba(OH)_2$ appears to be promising, serving a dual role as a sealer and hardener.

Clay or mud bricks: Most arid and semi-arid areas of the world permit building with unfired, sun-baked clay, a cheap and ever-available building material. Common admixtures of sand and cut straw to the clay lessen shrinkage during drying and increase the strength. In a study of prehistoric adobe buildings in the Middle East, TORRACA (1970) has observed that the greatest damage to ancient adobe ruins had occurred by undercutting of walls near the ground surface; damage to the top surfaces of the walls was also heavy. Detachment of the surface layer of the adobe bricks, mostly by ascending salts, is believed to be responsible for most of the damage. STEEN (1970) records similar decay of early Spanish adobe ruins in the American southwest. Rising saline ground moisture can be drained from the site through ditches; surface damage by the impact of driving rain may be checked by surface treatment. The use of animal blood is the oldest protective additive to adobe clay materials, both as an admixture and as a surface paint. Through laboratory experiments, WINKLER (1956) found that such treatment increased plasticity, strength and resistance of the clay to slaking in water. Kaolinitic and illitic clays responded most favorably to such treatment. A tight, monomolecular layer of blood around the tiny clay crystals is responsible for the high water repellence and the increased strength. Flash-dried, water-soluble mammal blood was suggested by WINKLER for underdeveloped countries, as this material appears to be a practical and inexpensive solution for the treatment of clay surfaces by spray or paint brush. Water-soluble dried blood is commercially used as a glue for plywood. The blood solution hardly changes the color of the clay surface nor does it reduce the porosity. Cloudbursts of short duration, characteristic of semi-arid areas, do little harm to treated adobe walls.

The Phoenicians painted their homes of adobe clay with bull's blood, the symbol of the "sap of life" to Mithraism, to protect the citizens' property from attack by Ahriman, the evil power. The ancient Hebrews painted the thresholds of their houses annually with animal blood at Passover time. Today the application of animal blood to clay is still practiced in semi-arid Eastern Europe where the religious meaning is now absent.

TORRACA (1970) experimented with the more expensive ethyl silicate, STEEN (1970) with a polymethane resin (trade name Pencapsula) developed by oil companies for the U.S. National Park Service as a hardener for adobe structures and for crumbling sandstone. The reports by both authors are promising.

13.2. Hardeners

Crumbling sandstone, weakened by weathering, should be impregnated with a hardener in order to prevent further disintegration. There must be sufficient stone porosity to maintain proper penetration of the hardening fluid throughout the stone. Sandstone is the most common stone subject to treatment. Sand-sized

Fig. 149. Dismantling of the statues of the Great Temple, Abu Simbel, Egypt, U.A.R. Crane lifts part of upper legs of figure on vertically mounted tie rods after careful separation by sawing. Photo by NENADOVIC, courtesy of UNESCO, Paris (Desroches-Noblecourt, 1968)

grains of generally round habit are held together by a variety of natural cements which may range from a mere contact film with much pore space left between the sand grains, to complete filling of all pores with cement. Weathering of sandstone consists of the solution and subsequent removal of the rock cement near the stone surface; the result is crumbling and finally sanding.

Silicification by treatment with organic silicates appears to be most appropriate because they form a natural and most efficient bond with the quartz grains of the sandstone. SCHMIDT (1969) reports 3- to 4-cm. penetration of an ethyl silicate solution in unweathered Baumberg sandstone, deeper into weathered stone. (This stone is widely used in the Rhein-Ruhr industrial area.) Recrystallization within the sandstone was so complete that no efflorescence or discoloring marred the surface appearance of the stone after treatment. Total removal of entrapped soluble salts by vacuum before treatment insures permanence of the impregnation.

Fig. 150. Low-pressure pump for the injection of low viscosity resins into crumbly Nubian sandstone, Abu Simbel, Upper Egypt. Photo from CARON (1967), courtesy of the author, Soletanche, Paris

Epoxy resins and monomers are often preferable to inorganic K-silicates because of their lower viscosity. FRANKLIN (1969) experimented with epoxy resins, unsaturated polyester resins and monomers as rock hardeners for the preparation of petrographic thin sections. Methyl methacrylate monomer hardens with benzoyl peroxide to clear plexiglass after three days at 40° C; styrene monomer hardens with benzoyl peroxide to clear, moderately hard polystyrene when cold.

The relocation of the massive statues at Abu Simbel, Upper Egypt, which were carved in crumbly, crossbedded Nubian sandstone more than 3000 years ago, required careful hardening of the stone before cutting and hoisting uphill away from flooding Lake Nasser above the Aswan High Dam. CARON (1967) describes the process of stone infiltration with epoxy resins and polyesters as applied in Abu Simbel (Fig. 150). Low-viscosity (250 centipoise) polyesters were preferred over epoxies with 1000 cP; these resins could be injected at temperatures from freezing to above 100° C. Additives lowered the viscosity to near 20 cP. Injection pressures of 5 to 6 kg./cm.² with borehole intervals of 20 to 30 cm. were most effective at Abu Simbel. Stone impregnation of the crumbling Nubian sandstone served three general purposes, i.e., rock consolidation, filling of open cracks and sealing of anchoring rods. The gigantic operation of relocation of the monument was recorded by DESROCHES-NOBLECOURT and GERSTNER (1968, Fig. 149).

13.3. Cleaning of Stone Surfaces

Washing of stone surfaces with water and detergents, and sandblasting have long been practiced for the purpose of removing disfiguring layers of soot and dust. Natural surface coatings developed by the prolonged exposure of the stone to its environment are often protective, but they may also be destructive; careful appraisal of the prevailing conditions should precede any treatment. STAMBOLOV (1970) cites successful sandblasting of buildings in England, whereby 15 mm. of rock substance was removed. Washing stone with water and detergents has often resulted in unexpected and undesireable migration of moisture through the stone. For general stone cleaning the following mixture has long been used successfully:

6 parts water, 1 part non-ionic detergent, 3 parts white spirit, 2 parts trichloro-ethylene.

The removal of iron and copper stains from silicate rocks can readily be performed with a 2% hydrofluoric acid solution.

The impregnation and cleaning of stone encounters a problem in the large heterogeneity of the rock substance and of rock properties, both physical and chemical; each stone can therefore be expected to respond to treatment differently. The travel routes of moisture and the supply of soluble salts within the building or monument add hidden variables to the already uncertain response of the rock substance. The great complexity of the interaction of chemicals with the stone substance does not permit a standardized treatment but requires the full understanding of the variables. This complexity of unknown factors may be compared with the understanding of the human mind which is superimposed on and influenced by the immediate environment. The "restorer" of a disturbed human mind, the psychiatrist, attempts to understand the variables and their interacting influences before proper and successful "repair" can be started. A close cooperation between the science disciplines can lead first to full understanding and then to successful treatment of a stone surface.

Summary

Modern impregnation treatment of stone usually serves a dual role, as an impregnator and as a hardener. Water-tight surface seals are no longer applied since moisture trapped inside the stone may do much damage. Impregnation treatment should be applied after removal of water-soluble salts. Table 38 summarizes a few common methods in use.

Surface cleaning is often done by sandblasting or by washing with 2% hydrofluoric acid. The stone response to any one chemical is very difficult to predict, as the interaction of unknown variables is complex. Unwise treatment can do more harm than good.

Table 38. *Stone Treatment by Chemicals*

Rock	Chemical	Interaction with Stone
Limestone, marble	sodium fluoride	forms insoluble CaF_2
	barium hydroxide	forms insoluble $BaCO_3$
Sandstone	potassium silicate	silica bond with quartz grains
	ethyl silicate	silica bond with quartz grains
	epoxy resins, polyesters	mechanical bond
Adobe clay	blood	water-repellent monomolecular layer around clay crystals
	ethyl silicate, polymethane resin	silica bond with clay crystals effect like blood

References

1. BURGESS, G., and R. J. SCHAFFER, 1952: Cleopatra's Needle. Chemistry and Industry, Oct. 18, 1952, pp. 1026—1029.
2. CARON, M. S., 1967: Applications des résines synthétiques dans les travaux publics. Annals de l'Institut Technique du Bâtiment et des Travaux Publics, 1976, No. 235—236, pp. 996—1016.
3. DESROCHES-NOBLECOURT, C., et G. GERSTER, 1968: Le Monde sauve Abou Simbel. Wien-Berlin: Verlag A. F. Koska, 273 p.
4. FRANKLIN, J. A., 1969: Rock impregnation trials using monomers, epoxide, and unsaturated polyester resins. J. Sedimentary Petrology, **39 (3)**, 1251—1253.
5. HEDVALL, J. A., 1966: Zerfall und Restaurierung von Kulturdenkmälern. Die Naturwissenschaften, **53 (19)**, 213—228.
6. HEDVALL, J. A., 1962: Chemie im Dienst der Archäologie, Bautechnik, Denkmalpflege. Akademiforlaget Gumperts, Göteborg 1962, 226 p.
7. HEMPEL, K. F. B., 1968: Notes on the conservation of sculpture, stone, marble and terracotta. Studies in Conservation, **13 (1)**, 34—44.
8. LEWIN, S., 1966/67: The conservation of limestone objects and structures. Conferences on the weathering of stones; Council of Monuments and Sites. Icomos, Paris 1968, pp. 41—64.
9. MUNNIKENDAM, R. A., 1967: Preliminary notes on the consolidation of porous building materials by impregnation with monomers. Studies in Conservation, **12 (4)**, 158—162.
10. MUNNIKENDAM, R. A., and T. J. WOLSCHRIJN, 1969: Further remarks on the impregnation of porous materials with monomers. Studies in Conservation, **14**, 133—135.
11. PLENDERLEITH, H. J., 1956: The conservation of antiquities and works of art. London: Oxford University Press, 375 p.
12. RIEDERER, J., 1970: Stone preservation in Germany. Conservation of Stone and Wooden Objects, 7—13 June, 1970, New York, Conference Proceedings, 176 Old Brompton Rd., London S.W. 5, pp. 125—133.
13. SCHMIDT-THOMSEN, K., 1969: Zum Problem der Steinzerstörung und Konservierung. Deutsche Kunst- und Denkmalpflege, pp. 11—23.

14. STAMBOLOV, T., 1970: Conservation of Stone. Conservation of Stone and Wooden Objects, 7—13 June, 1970, New York, Conference Proceedings, 176 Old Brompton Rd., London S.W. 5, pp. 119—123.

15. STAMBOLOV, T., 1968: Notes on the removal of iron stains from calcareous stone. Studies in Conservation, **13** (**1**), 45—47.

16. STEEN, C. R., 1970: Some recent experiments in stabilizing adobe and stone. Conservation of Stone and Wooden Objects, 7—13 June, 1970, New York, Conference Proceedings, 176 Old Brompton Rd., London S.W. 5, pp. 59—64.

17. TORRACA, G., 1970: An international project for the study of mud-brick preservation. Conservation of Stone and Wooden Objects, 7—13 June, 1970, New York, Conference Proceedings, 176 Old Brompton Rd., London S.W. 5, pp. 47—58.

18. WINKLER, E. M., 1956: The effect of blood on clays. Soil Science, **82** (**2**), 157—164.

19. WINKLER, E. M., 1961: Method of stabilizing soil with soluble dried blood. U.S. Patent No. 2,971.293, Febr. 14, 1961, Washington, D.C.

Appendix A

Properties of Some Rock Forming Minerals

Mineral	Mohs Hardness	Specific Gravity	Color	Cleavage Directions	Vol. Expn., % 20° C to:			Weathering Product or Solubility	Occurrence	Other Properties
					200°	500°	600°			
Elements:										
Graphite C	1—2	2.2	black	basal				flaky residue in soils	metam. rocks: marble, slate schist	
Oxides:										
Quartz SiO$_2$ room temp. above 573°	7	2.65 2.40	white	none	.75	2.75	4.5		granite, sandstones quartzite	silicosis; lack of fire resistance
Chert	7	2.65	white	none	same				limestones, dolomites	crypto-crystal-line chert and
Opal	5—6	1.9—2.2	white						sediments, weathered igneous rocks	amorphous opal expand in concrete
Hematite Fe$_2$O$_3$	5½—6½	5.16	red, black	none	.5	1.5		limonite, ochre	pigment in sediments	hydrates slowly to limonite, rust burst
Magnetite Fe$_3$O$_4$	5½	5.2	black	none	.5	1.9	2.3	limonite, ochre	igneous rocks metamorphic	weathers very slowly to limonite
Goethite α-FeOOH	5.0—5½	3.6—4.0	brown, ochre	none				weathering end product	sediments soils	

Mineral	Hardness	Density	Color	Cleavage				Weathering end product	Occurrence	Remarks
Lepidocrocite γ-FeOOH	5	4.09	red, orange	none					sediments soils	rare in nature
Limonite $Fe_2O_3 \cdot nH_2O$			brown, ochre	none				like goethite	sediments soils	limonite is goethite plus amorphous ferric hydroxide
Sulfides:										
Pyrite FeS_2	6.0–6½	5.0–5.2	yellow	none				limonite	common in all rocks	SO_2 ion released which may oxidize to SO_3.
Marcasite FeS_2	6.0–6½	4.8–4.9	yellow	none				limonite	sediments	Expands during oxidation and hydration to limonite
Carbonates:										
Calcite $CaCO_3$	3	2.72	white	3, not at right angles	.25	1.2	1.9	0.92 g./l. at 20°C at a $P_{CO_2} = 0.034\%$	limestones, dolomites, marbles	solubility of all carbonates rises rapidly with increasing P_{CO_2} in industrial and urban atmospheres at low temperatures
Dolomite $CaMg\,(CO_3)_2$	3½–4.0	2.85	white	same as calcite				similar to calcite	dolomites, marbles	
Magnesite $MgCO_3$	4.0–4½	3.0	white	same as calcite				similar to calcite	serpentines	
Siderite $FeCO_3$	4–4½	3.89	grey	same as calcite				ca. 0.50 g./l. at 20°C at a $P_{CO_2} = 0.034$	limestones, marbles, schist	

Mineral	Mohs Hardness	Specific Gravity	Color	Cleavage Directions	Vol. Expn., % 20° C to: 200°	500°	600°	Weathering Product or Solubility	Occurrence	Other Properties
(Thermonatrite) $Na_2CO_3 \cdot H_2O$	1–1½	2.26	white	1				soluble	desert floor stone pores	important in salt bursting
(Natron) $Na_2CO_3 \cdot 10\,H_2O$	1–1½	1.48	white	2				soluble	desert floor stone pores	important in salt bursting
Sulfates:										
Anhydrite $CaSO_4$	3	2.95	white	3				0.2—0.5 g./l.	sediments desert floor	common in efflorescence
Gypsum $CaSO_4 \cdot 2\,H_2O$	2	2.32	white	3				1.75—2.15 g./l.	sediments desert floor	
(Kieserite) $MgSO_4 \cdot H_2O$	3½	2.57	white	4				slow solution	evaporites desert floor in stone	
(Epsomite) $MgSO_4 \cdot 7\,H_2O$	2—2½	1.68	white	2				slow solution	evaporites in stone	
Silicates: Feldspars:										
Orthoclase $KAlSi_3O_8$	6	2.57	white, red	2	0.15	0.75	1.1	kaolinite illite		all feldspars are good indicators of the rock freshness
Albite $NaAlSi_3O_8$	6	2.60	white	2	0.45	1.2	1.45	kaolinite illite	igneous metamorphic	
Labradorite 60% $[CaAl_2Si_2O_8]$ 40% $[NaAlSi_3O_8]$	6	2.70	dark	2	0.3	0.66	0.75	kaolinite illite		

Mineral	Hardness	Density	Color	Cleavage				Weathering product	Occurrence	Remarks
Muscovite $KAl_2(AlSi_3O_{10}) \cdot (OH)_2$	2 – 2½	2.80	white	1					igneous sediments metamorphic	very resistant to weathering
Biotite $K(Mg, Fe)_3 \cdot (AlSi_3O_{10})(OH)_2$	2½ – 3	3.1	black	1				loss of Fe, Mg	igneous sediments metamorphic	rusty dicoloration by loss of Fe
Kaolinite $Al_2O_3 \cdot 2 SiO_2 \cdot 4 H_2O$	2	2.6	white	1				end product of weathering	sediments	absorbs water without swelling
Illite $KAl_2(AlSi_3O_{10}) \cdot (OH)_2$	2½	2.8	white	1				end product	sediments	"clay-mica", similar to kaolinite
Montmorillonite $Al_2O_3 \cdot 4 SiO_2 \cdot n H_2O$	2.5	2 – 2.7	white	1				end product	sediments	swelling clay; swelling pressure to 9000 psi
Glauconite $(K, Ca, Na), (Al, Fe'', Fe''', Mg)_2 ([OH]_2Si_3, AlO_{10})$	—	2.2 – 2.8	green	1					sediments	important green pigment
Hornblende Fe-Mg silicate	5 – 6	3.0 – 3.4	green, black	2				brown clay, limonite	igneous metamorphic	weathers slowly; gives strength and toughness to rocks
Augite Fe-Mg silicate	5½ – 6	3.3 – 3.5	brown, green, black	2	0.45	1.27	1.57	brown clay, limonite	igneous metamorphic	to rocks

Appendix B

Stone Specifications (ASTM)

ASTM (American Society for Testing and Materials) stone specifications will be presented in abridged form, for granite, sandstone, limestone-marble, slate, and concrete aggregates. The specifications are suggested guide lines for stone for a given application. Almost all industrialized countries have similar standards, which may be readily obtained from the agency concerned with building codes.

1. Structural Granite (C 615—68)

Structural granite shall include all varieties of commercial granite that are sawed, cut, split, or otherwise shaped for building purposes. Structural granite shall be classified:

I. *Engineering Grade:* bridge piers, sea and river walls, dams, and related structures, bridge superstructures, grade separations, and retaining walls, flexural members, curbstone and pavements.

II. *Architectural Grade:* Monumental structures, institutional buildings, commercial buildings, residential buildings, landscaping, parks, and other ornamental and private improvements.

Physical requirements: Structural granite shall conform to the physical requirements prescribed in Tables 1 and 2. Structural granite shall be sound,

Table 1. *Physical Requirements of Structural Granite*

Physical Property	Test Requirement	ASTM Test
Absorption by weight, max., per cent	0.4	C 97
Density, min., lb./ft.^{3} (kg./m.^{3})	160 (2560)	C 97
Compressive strength, min., psi (kg./mm.^{2})	19,000 (13.4)	C 170
Modulus of Rupture, min., psi (kg./mm.^{2})	1500 (1.05)	C 99

durable, and free from imperfections such as starts, cracks and seams that would impair its structural integrity. Granite shall be free from minerals that will cause objectionable staining. The desired color and the permissible natural variations in color and texture shall be specified by carefully detailed description.

Table 2. *Physical Requirements for Different Life Expectancies*

Specific Use	Life Expectancy	
	less than 50 yr.	more than 50 yr.
	Compr. Strength min., psi	Compr. Strength min., psi
Engineering Grade:		
Bridge piers, sea and river walls, dams	25,000	30,000
Bridge superstructures, grade separations and retaining walls	25,000	30,000
Flexural members (modulus of rupture not less than 2000 psi)	30,000	30,000
Traffic controls, etc.	25,000	30,000
Architectural Grade:		
Monumental buildings	28,000	30,000
Institutional buildings	26,000	28,000
Commercial buildings	20,000	26,000
Residential buildings	16,000	20,000
Landscaping, parks, etc.	25,000	30,000

From: ASTM Designation C 422-58 T

2. Building Sandstone (C 616—68)

The sandstone shall be free from seams, cracks, or other imperfections that would impair its structural integrity. The color desired and the permissible natural variations in color and texture shall be specified in careful detail. Sandstone containing minerals such as pyrite and marcasite, that may upon exposure cause objectionable stain, shall be excluded. Table 3 lists additional requirements.

Table 3. *Physical Requirements of Building Sandstone*

Property	Sandstone	Quartzitic Sandstone	Quartzite	ASTM Test
Free silica content, min., %	60	90	95	
Absorption, max., %	20	3	1	C 97
Compressive strength, taken in weakest direction, min., psi (kg./mm.²)	2000 (1.4)	10,000 (7.0)	20,000 (14.1)	C 170
Modulus of rupture, min., psi (kg./mm.²	300 (0.2)	1000 (.7)	2000 (1.4)	C 99

3. Dimension Limestone (C 563—67)

Dimension limestone shall include stone that is sawed, cut, split, or otherwise finished or shaped, and shall specifically exclude molded, cast, or otherwise artificially aggregated units composed of fragments, and also crushed and broken stone. Dimension limestone may be classified into three categories, generally descriptive of those limestones having densities in approximate ranges:

I. (Low-density) Limestone with density ranging from 110 through 135 lb./ft.^{3} (1.76 through 2.16 g./cm.^{3}).

II. (Medium-density) Limestone with density greater than 135 and not greater than 160 lb./ft.^{3} (2.16 through 2.56 g./cm.^{3}).

III. (High-density) Limestone with density greater than 160 lb./ft.^{3} (2.56g./cm.^{3}).

Physical requirements: Dimension limestone shall be sound, durable, and free from visible defects or concentrations of materials that will cause objectionable staining or weakening under normal environments of use. Additional specifications are given in Table 4.

Table 4. *Physical Characteristics of Dimension Limestone*

Categories	Absorption, max., %	Compr. Strength min., psi (kg./mm.^{2})	Modulus Rupture min., psi (kg./mm.^{2})
I (low-density)	12	1800 (1.25)	400 (0.28)
II (medium-density)	7.5	4000 (2.8)	500 (0.35)
III (high-density)	3	8000 (5.6)	1000 (.70)

4. Exterior Marble (C 503—67)

Marble is a crystalline rock composed predominantly of one or more of the following minerals: calcite, dolomite, or serpentine, and capable of taking a polish.

Physical requirements: For exterior use, marble shall be sound, free from spalls, cracks, open seams, pits, or other defects that would impair its strength, durability, or appearance. See also Table 5.

Table 5. *Physical Requirements of Exterior Marble*

Physical Property	Requirement	ASTM Test
Absorption, max.	0.75%	C 97
Specific gravity, min.		
calcite	2.60	C 97
dolomite	2.80	C 97
serpentine	2.70	C 97
travertine	2.30	C 97
Compressive strength, min., psi (kg./mm.^{2})	7500 (5.27)	C 170
Modulus of rupture, min., psi (kg./mm.^{2})	1000 (0.70)	C 99
Abrasion resistance, min., H$_a$	10.0	C 241

5. Roofing Slate (C 406—58)

For natural slate shingles as commonly used on sloping roofs and also square or rectangular tiles for flat roof coverings.

Physical requirements: Three grades are covered, based on the length of service that may be expected. Details are given in Table 6.

Slates are manufactured for standard roofs, for textural roofs, for graduated roofs, and for flat roofs. The slate color shall agree closely with that of accepted samples; the following color nomenclature is used: black, blue black, gray, blue gray, purple, mottled purple and green, green, purple variegated, weathering green (changes to buff and brown).

Imperfections: Curvature shall not exceed 1/8″ in 12″.

Knots and knurls are not objectionable on the top face. Slate may be rejected if the protuberances project more than 1/16″ beyond the split surface.

Ribbons: Grades S_1 and S_2 shall be free from soft ribbons, S_3 shall be free of soft ribbons below both nail holes.

Table 6. *Physical Requirements of Roofing Slate*

Designation	Service Period, years	Modulus of Rupture across grain, min., psi	Absorption max., %	Depth of Softening max., in.
Grade S_1	75 to 100	9000	0.25	0.002
Grade S_2	40 to 75	9000	0.36	0.008
Grade S_3	20 to 40	9000	0.45	0.014

6. Structural Slate (C 629—68)

For general building and structural purposes.

General requirements: Slate shall be sound, free from spalls, pits, cracks or other defects. In general, slate for exterior application in ambient acidic atmosphere or in industrial areas where heavy air pollution occurs shall be free from carbonaceous ribbons. Abrasion hardness requirements pertain to slate subject to foot traffic only. Further requirements are given in Table 7.

Table 7. *Physical Requirements for Structural Slate*

Physical Property	Exterior Use	Interior Use	ASTM
Absorption, max., %	0.25	0.45	C 121
Modulus of Rupture, min., psi (kg./mm.^{2})			
across grain	9000 (6.3)	9000 (6.3)	C 120
along grain	7200 (5.0)	7200 (5.0)	C 120
Abrasion Hardness, min, H_a	8.0	8.0	C 241
Acid Resistance, max., in. (mm.)	0.015 (0.38)	0.025 (0.64)	C 217

7. Concrete Aggregates (C 33—67)

These specifications cover fine and coarse aggregate, other than lightweight aggregate, for use in concrete. Separate treatment is given to fine and coarse aggregate. The sieve analyses are not given here. Tables 8 and 9 added limitations.

Table 8. *Limits for Deleterious Substances*

Item	Maximum Per cent by Weight of Total Sample		ASTM
	Fine Aggregate	Coarse Aggregate	
Friable Particles	1.0	0.25	C 142
Soft Particles	—	5.0	C 235
Coal and Lignite	0.5 to 1.0	0.5 to 1.0	C 123
Material finer than No. 200 sieve			C 117
subject to abrasion	3.0		C 125
all other concrete	5.0	1.0	
Chert as an impurity*			
severe exposure		1.0	
mild exposure		5.0	

   * Chert as an impurity that will disintegrate in 5 cycles of the soundness test, or 50 cycles of freezing and thawing in water, or that has a specific gravity, saturated-surface dry, of less than 2.35.

Table 9. *Physical Properties of Coarse Aggregate for Concrete*

Physical Property	Gravel, Crushed Gravel, or Crushed Stone	ASTM
Soundness, loss in five cycles, max., % by weight		
sodium sulfate	12	C 88
magnesium sulfate	18	C 88
Abrasion, max. loss, % by weight	50	C 131

Appendix C

Conversion Tables

1. Dimensions

from	to	multiply by	from	to	multiply by
inch	centimeter	2.54	centimeter	feet	0.0328
inch	feet	0.0833	centimeter	inch	0.3937
inch	meter	0.0254	centimeter	meter	0.01
inch	millimeter	25.4	centimeter	micron	1,000
inch	micron	25.4×10^3	centimeter	millimicron	1×10^7
feet	centimeter	30.48	millimeter	Ångstrom	1×10^7
feet	inch	12.00	millimeter	centimeter	0.1
feet	meter	0.3048	millimeter	feet	0.0032
micron	centimeter	0.0001	millimeter	inch	0.0393
micron	inch	3.9370×10^{-5}	millimeter	micron	1000
micron	millimeter	1000			
micron	millimicron	1000			

2. Area

from	to	multiply by
square centimeter	square feet	0.00107
square centimeter	square inch	0.1550
square centimeter	square meter	0.0001
square centimeter	square millimeter	100
square millimeter	square centimeter	0.01
square millimeter	square inch	0.00155
square feet	square meter	0.0929
square inch	square centimeter	6.4516
square inch	square feet	0.0069
square inch	square millimeter	645.16

3. Volume

from	to	multiply by
pounds/cubic foot	gram/cubic centimeter	0.01601
pounds/cubic inch	gram/cubic centimeter	27.6799
pounds/cubic inch	gram/liter	27.6806
pound/gallon (U.S.)	gram/cubic centimeter	0.01198
pound/gallon (U.S.)	pound/cubic foot	7.4805
gram/cubic centimeter	gram/millimeter	1.0000
gram/cubic centimeter	pound/cubic foot	62.4279
gram/cubic centimeter	pound/cubic inch	0.0361
gram/liter	part per million (ppm)	1000
gram/milliliter	gram/cubic centimeter	0.9999
milligram/liter	part per million (ppm)	1.000

4. Pressure

from	to	multiply by
atmosphere	bar	1.0132
atmosphere	kilogram/centimeter square	1.0332
atmosphere	millimeter Hg/0° C	760
atmosphere	pound/square inch	14.6960
kilogram/square centimeter	atmosphere	0.9678
kilogram/square centimeter	pound/square inch	14.2233
kilogram/square millimeter	pound/square inch	1422.3343
pound/square inch	atmospere	0.0680
pound/square inch	gram/square centimeter	70.3069
pound/square inch	kilogram/square centimeter	0.0703
bar	atmosphere	0.9869
bar	gram/square centimeter	1019.716
bar	kilogram/square centimeter	1.0197
bar	pound/square inch	14.5038

5. Temperature

from	to	multiply by
degrees Centigrade	degrees Fahrenheit	5/9 (° F) -32
degrees Fahrenheit	degrees Centigrade	9/5 (° C) $+32$
degrees Kelvin	degrees Centigrade	273.15 $+$° C
degrees Centigrade	degrees Kelvin	273.15 $-$° C

6. Heat Transmission

1 cal per centimeter/second/square centimeter/degree Centigrade
    = 2903 Btu per inch/hour/square foot/degree Fahrenheit
    = 241.9 Btu per foot/hour square foot/degree Fahrenheit

1 Btu per inch/hour/square foot/degree Fahrenheit
    = 0.0003445 cal centimeter/second/square centimeter/degree Centigrade
    = 0.08333 Btu foot/hour/square foot/degree Fahrenheit

Appendix D

Glossary of Geological and Technical Terms, Exclusive of Minerals and Architectural Terms

Absorption. Taking up, assimilation, or incorporation of liquids in a solid (AGI).

Adsorption. Adhesion of molecules of gases or molecules in solution to the surfaces of solid bodies with which they are in contact (AGI).

Agate. A variegated waxy quartz in which the colors are in bands, in clouds or in distinct groups (AGI). Term is frequently (erroneously) applied to quartz in granite or veins with bluish waxy appearance.

Aging. Field storage of stone block for the purpose of drying, stress relief, and case hardening.

Aphanitic. Pertaining to a texture of rocks in which the crystalline constituents are too small to be distinguished with the unaided eye. It includes both microcrystalline and cryptocrystalline textures (AGI).

Aplite. A dike rock consisting almost entirely of light-colored mineral constituents and having a characteristic fine-grained granitic texture. Aplites may range in composition from granitic to gabbroic, but when the term is used with no modifier it is generally understood to be granitic, i.e., consisting essentially of quartz and orthoclase (AGI).

Argillite. A rock either derived from siltstone, claystone, or shale, that has undergone a somewhat higher degree of induration that is present in those rocks with an intermediate position to slate. Cleavage is approximately parallel to bedding in which it differs from slate (AGI).

Arkose. A rock of granular texture, formed principally by process of mechanical aggregation. It is essentially composed of of large grains of clear quartz and grains of feldspar, either lamellar or compact. These two minerals are often mixed in almost equal quantities, but oftener quartz is dominant (AGI).

Arkosic Limestone. An impure clastic limestone containing a relatively high proportion of grains and/or crystals of feldspar, either detrital or formed in place (AGI).

Arkosic Sandstone. A sandstone in which much feldspar is present. This may range from unassorted products of granular disintegration of fine or medium-grained granite to a partly sorted river-laid or even marine arkosic sandstone (AGI).

Ashlar. Rectangular blocks having sawed, planed, or rock-faced surfaces, contrasted with cut blocks which are accurately sized and surface tooled. May be laid in courses (Stone Catalog).

Band. A term applied to a stratum or lamina conspicuous because it differs in color from adjacent layers; a group of layers displaying color differences is described as being banded (AGI).

Bed. 1. In granites and marbles a layer or sheet of the rock mass that is horizontal, commonly curved and lenticular, as developed by fractures. Sometimes also applied to the surface of parting between sheets.
2. In stratified rocks the unit layer formed by sedimentation, of variable thickness, and commonly tilted or distorted by subsequent deformation; generally develops a rock cleavage, parting, or jointing along the planes of stratification (Stone Catalog).

Bedding Plane. In sedimentary or stratified rock, the division planes which separate the individual layers, beds or strata (AGI).

Bluestone. The commercial name for a dark bluish-gray feldspathic sandstone or arkose. The color is due to the presence of fine black and dark-green minerals, chiefly hornblende and chlorite. The rock is extensively quarried in New York. Its toughness, due to light metamorphism, and the ease with which it may be split into thin slabs especially adapt it for use as flagstone. The term has been locally applied to other rocks, among which are dark-blue slate and blue limestone (AGI).

Breccia. A fragmental rock whose components are angular and therefore, as distinguished from conglomerates, are not waterworn. There are friction or fault breccias, talus-breccias and eruptive (volcanic) breccias. The word is of Italian origin (AGI).

Breccia Marble. Any marble made up of angular fragments (AGI).

Breccia Vein. A fissure filled with fragments of rock in the interstices of which vein matter is deposited (AGI).

Broach. To drill or cut out material left between closely spaced drill holes (Stone Catalog).

Brownstone. Ferruginous sandstone in which the grains are generally coated with iron oxide. Applied almost exclusively to a dark brown sandstone derived from the Triassic of the Connecticut River Valley (AGI).

Brushed Finish. Obtained by brushing the stone with a coarse rotary-type wire brush (Stone Catalog).

Burst (or Bump). See Rock Burst.

Calcite Limestone. A limestone containing not more than 5% of magnesium carbonate (ASTM).

Calcite Marble. A crystalline variety of limestone containing not more than 5% of magnesium carbonate (ASTM).

Calcite Streaks. Description of a white or milky-like streak occurring in stone. It is a joint plane usually wider than a glass seam and has been recemented by deposition of calcite in the crack and is structurally sound (Stone Catalog).

Cement. Chemically precipitated material occurring in the interstices between allogenic particles of clastic rocks. Silica, carbonates, iron oxides and hydroxides, gypsum and barite are the most common. Clay minerals and other fine clastic particles should not be considered cement (AGI).

Chert. A compact, siliceous rock formed of chalcedonic or opaline silica, one or both, and of organic or precipitated origin. Chert occurs distributed through limestone, affording chert limestone. Petrographically, chert is composed of microscopic chalcedony, quartz particles or both whose outlines range from easily resolvable to nonresolvable with the stereoscopic microscope. Particles rarely exceed 0.5 mm. (AGI).

Chroma. The chroma notation of a color indicates the strength, the saturation, or the degree of departure of a particular hue from a neutral grey of the same value. The scales of chroma extend from /0 for a neutral grey, to /10, /12, /14, depending upon the strength or saturation of the individual color.

Clastic. A textural term applied to rocks composed of fragmental material derived from pre-existing rocks or from the dispersed consolidation products of magmas or lavas. The commonest clastic rocks are sandstones and shales as distinct from limestone and anhydrites (AGI).

Cleavage. 1. The tendency of a rock to cleave or split along definite, parallel, closely spaced planes which may be highly inclined to bedding planes. It is a secondary structure, commonly confined to bedded rocks, is developed by pressure, and ordinarily is accompanied by some recrystallization of the rocks (AGI).

2. In the stone industry, cleavage is the ability of a rock to break along natural surfaces, a surface of natural parting (Stone Catalog).

Cobblestone. A natural rounded stone, large enough for use in paving (Stone Catalog).

Commercial Marble. A crystalline rock composed predominantly of one or more of the following minerals: calcite, dolomite or serpentine, and capable of taking a polish (ASTM).

Conglomerate. A rock made up of worn and rounded pebbles of various sizes cemented as in sandstone. It includes varieties locally known as "pudding-stone" (ASTM).

Crossbedding. An original lamination oblique to the main stratification. Lamination, in sedimentary rocks, confined to single beds and inclined to the general stratification (AGI).

Crowfeet. (Stylolite-)Description of a dark gray to black zigzag marking occurring in stone. Usually structurally sound (Stone Catalog).

Crystalline Marble. A limestone, either calcitic or dolomitic, composed of interlocking crystalline grains of the constituent minerals, and of phaneritic texture. Commonly used synonymously with "marble", and thus representing a recrystallized limestone. Improperly applied to limestones that display some obviously crystalline grains in a fine-grained mass but which are not of interlocking texture and do not compose the entire mass (Stone Catalog).

Cut Stone. This includes all stone cut or machined to given sizes, dimension or shape, and produced in accordance with working or shop drawings which have been developed from the architect's structural drawings (Stone Catalog).

Cutting Stock. A term used to describe slabs of varying size, finish, and thickness which are used in fabricating treads, risers, copings, borders, sills, stools, hearths, mantels and other special purpose stones (Stone Catalog).

Dendrites. A branching figure resembling a shrub or tree, produced on or in a mineral or rock by the crystallization of a foreign mineral, usually an oxide of manganese, as in moss agate; also the mineral or rock so marked (AGI).

Diabase. A rock of basaltic composition, consisting essentially of labradorite and pyroxene, and characterized by ophitic texture (discrete crystals or grains of pyroxene fill the interstices between lath-shaped feldspar crystals (AGI).

Dimension Stone. Stone pre-cut and shaped to dimensions of specified sizes (Stone Catalog).

Diorite. An igneous rock composed essentially of sodic plagioclase and hornblende, biotite, or pyroxene. Small amounts of quartz and orthoclase may be present (AGI).

Dissolution. The process of dissolving (AGI).

Dolomite. A limestone containing in excess of 40% magnesium carbonate as the dolomite molecule (ASTM).

Dolomitic Limestone. See Magnesian Limestone.

Dressed or Hand-Dressed Stone. Rough chunks of stone cut by hand to create a square or rectangular shape. A stone which is sold as dressed stone generally refers to stone ready for installation (Stone Catalog).

Dry Joint. An open or unhealed joint plane not filled with calcite and not structurally sound (Stone Catalog).

Efflorescence. A crystalline deposit appearing on stone surfaces caused by soluble salts carried through or onto the stone by moisture, which has sometimes been found to come from brick, tile, concrete blocks, mortar, concrete and similar materials in the wall or above (Stone Catalog).

Exposed Aggregate. The larger pieces of stone aggregate purposefully exposed for their color and texture in a cast slab (Stone Catalog).

Fabric. The orientation in space of the elements of which a rock is composed (AGI).

Fault. A fracture or fracture zone along which there has been displacement of the two sides relative to one another parallel to the fracture. The displacement may be a few inches or many miles (AGI).

Fault Fissure. The fissure produced by a fault, even though it is afterward filled by a deposit of minerals (AGI).

Fault Rock. The crushed rock due to the friction of the two walls of a fault rubbing against each other (AGI).

Fault Strike. The direction of the intersection of the fault surface, or the shear zone, with a horizontal plane (AGI).

Felsite. An igneous rock with or without phenocrysts, in which either the whole or the groundmass consists of a cryptocrystalline aggregate of felsic minerals, quartz and potassium feldspar being those characteristically developed. When phenocrysts of quartz are present the rock is termed a quartz felsite, or, more commonly, a quartz porphyry (AGI).

Field Stone. Loose blocks separated from ledges by natural processes and scattered through or upon the regolith (soil) cover; applied also to similar transported materials, such as glacial boulders and cobbles (Stone Catalog).

Flagstone. Thin slabs of stone used for flagging or paving walks, driveways patios, etc. It is generally fine-grained sandstone, bluestone, quartzite or slate, but thin slabs of other stones may be used (Stone Catalog).

Flat Joints. See Sheeting.

Flint. A dense, finegrained form of silica which is very tough and breaks with a conchoidal fracture and cutting edges. Of various colors, white yellow, gray, black. See also Chert (AGI).

Foliation. The laminated structure resulting from segregation of different minerals into layers parallel to the schistosity. Foliation is considered synonymous with slaty cleavage (AGI).

Frame Weathering. Transport of dissolved salts from the inside of a stone block towards the surface where mineral matter is redeposited near the surface of stone while the inside crumbles, leaving temporarily a frame-like structure.

Freestone. A sandstone which breaks easily in any direction without fracture or splitting (Stone Catalog).

Gabbro. A granular igneous rock rich in calcic plagioclase feldspar and hornblende or pyroxene. Dark in color, it is often marketed as Black Granite (ASTM).

Gang Sawed. Description of the granular surface of stone resulting from gang sawing alone (Stone Catalog).

Glass Seam. A joint plane in a rock that has been recemented by deposition of calcite or silica in the crack and is structurally sound (AGI).

Gneiss. A foliated crystalline rock composed essentially of silicate minerals with interlocking and visibly granular texture, and in which the foliation is due primarily to alternating layers, regular or irregular, of contrasting mineralogic composition. In general a gneiss is characterized by relatively thick layers as compared with a schist. According to their mineralogic compositions gneisses may correspond to other rocks of crystalline, visibly granular, and interlocking texture, such as those under definition of commercial granite, and may then be known as granite-gneiss if strongly foliated, or gneissic granite if weakly foliated (ASTM).

Gneissic Granite. Weakly foliated granite. See Granite.

Grain. 1. The particles or discrete crystals which comprise a rock or sediment.
2. A direction of splitting in rock, less pronounced than the rift and usually at right angle to it (AGI).

Granite. A true granite is a visibly granular, crystalline rock of predominantly interlocking texture, composed essentially of alkalic feldspars and quartz. Feldspar is generally present in excess of quartz, and accessory minerals (chiefly micas, hornblende, or more rarely pyroxene). The alkalic feldspar may be present as individual mineral species, as isomorphous or mechanical intergrowth with each other, or as chemical intergrowth with the lime feldspar molecule, but 80% of the feldspar must be composed of the potash or soda feldspar molecules (ASTM).

Granite-Gneiss. A gneiss visibly derived from granite by metamorphism. See Granite.

Greenstone. Includes rocks that have been metamorphosed or otherwise so altered that they have assumed a distinctive greenish color owing to the presence of one or more of the following minerals: chlorite, epidote, or actinolite (ASTM).

Groundmass. The material between the phenocrysts in porphyritic igneous rock. It includes the basis or base as well as the smaller crystals of the rock. Essentially synonymous with matrix (AGI).

Grout. Mortar of pouring consistency (Stone Catalog).

Hard Rock. Rock which requires drilling and blasting for its economical removal (AGI).

Hardway. In quarrying, especially in the quarrying of granite. The rift is the direction of easiest parting, the grain is a second direction of parting, and the hardway is the third and most difficult direction along which parting takes place (AGI).

Head. The end of a stone which has been tooled to match the face of the stone. Heads are used at outside corners, windows, door jams or any place where the veneering will be visible from the side (Stone Catalog).

Honed Finish. Honed is a superfine smooth finish (Stone Catalog).

Honeycomb Weathering. Differential weathering of sandstone by minor differences in the resistance to weathering also influenced by micro-climates.

Hue. Chromatic colors in the Munsell System of Color Notations are divided into five principal classes which are given the hue names of red (R), yellow (Y), green (G), blue (B), and purple (P). The hues extend around a horizontal color sphere about a neutral or a chromatic vertical axis.

Igneous. Formed by solidification from a molten or partially molten state; one of the two great classes into which all rocks are divided, and contrasted with sedimentary. Rocks formed in this manner have also been called plutonic rocks, and are often divided for convenience into plutonic and volcanic rock, but there is no clear line between the two (AGI).

Igneous Rock Series. An assemblage of igneous rocks in a single district and belonging to a single period of igneous activity, characterized by a certain community of chemical, mineralogical, and occasionally also textural properties (AGI).

Joint. 1. In geology, a fracture or parting which interrupts abruptly the physical continuity of a rock mass (AGI).

2. The space between stone units, usually filled with mortar. Joints are classified as: Flush, Rake, Cove, Weathered, Bead, Stripped, "V" (Stone Catalog).

Joint Set. A group of more or less parallel joints (AGI).

Joint System. Consists of two or more joint sets or any group of joints with a characteristic pattern, such as a radiating pattern, a concentric pattern, etc. (AGI). For classification see Natural Deformation of Rocks.

Lamination. The layering or bedding less than 1 cm. in thickness in a sedimentary rock (AGI).

Lava. Fluid rock such as that which issues from a volcano or a fissure in the earth's surface; also the same material solidified by cooling (AGI).

Liesegang Banding. Banding in color by diffusion (AGI).

Liesegang Rings. Rings or bands resulting from rhythmic precipitation in a gel (AGI).

Limestone. A rock of sedimentary origin (including chemically precipitated material) composed principally of calcium carbonate or the double carbonate of calcium and magnesium (ASTM).

Limestone Marble. Recrystallized limestones and compact, dense, relatively pure microcrystalline varieties that are capable of taking a polish are included in commercial marbles (ASTM).

Lineation. Linear parallelism or linear structure. The parallel orientation of structural features that are lines rather than planes. Lineation may be expressed by the parallel orientation of the following: long dimensions of minerals, striae on slickensides, streaks of minerals, cleavage-bedding intersection, intersection of two cleavages and fold axis (AGI).

Lithification. The complex of processes that converts a newly deposited sediment into an indurated rock. It may occur shortly after deposition (AGI).

Machine Finish. The generally recognized standard machine finish produced by the planers (Stone Catalog).

Magmatic Rock. See Igneous Rock.

Magnesian (Dolomitic) Limestone. A limestone containing not less than 5% nor more than 40% of magnesium carbonate (ASTM).

Magnesian (Dolomitic) Marble. A crystalline variety of limestone containing not less than 5 nor more than 40% of magnesium carbonate as the dolomite molecule (ASTM).

Marble (Scientific Definition). A metamorphic (recrystallized) limestone composed predominantly of crystalline grains of calcite or dolomite or both, having interlocking or mosaic texture. Compare with Commercial Marble (ASTM).

Masonry. Built-up construction, usually of a combination of materials set in mortar (Stone Catalog).

Massive. 1. Of homogeneous structure, without stratification, flow-banding, foliation, schistosity, and the like: often, but incorrectly, used as synonymous with igneous and eruptive.
2. Occurring in thick beds, free from minor joints and lamination (applied to some sedimentary rocks).

Metamorphic Rock. This group includes all those rocks which have formed in the solid state in response to pronounced changes of temperature, pressure, and chemical environment, which take place, in general, below the shells of weathering and cementation (AGI).

Milky. Fractured quartz and microcrystalline quartzites may have milky opalescent surfaces. Thick chunks of vein quartz from pegmatites constitute a popular feature stone for fire places, etc.

Monzonite. A granular plutonic rock containing approximately equal amounts of orthoclase and plagioclase, and thus intermediate between syenite and diorite. Quartz is usually present, but if it exceeds 2% by volume the rock is classified as quartz monzonite (AGI).

Natural Bed. The setting of the stone on the same plane as it was formed in the ground. This generally applies to all stratified materials (Stone Catalog).

Natural Cleft. When stones formed in layers are cleaved or separated along a natural seam the remaining surface is referred to as a natural cleft surface (Stone Catalog).

Nicked-Bit Finish. Obtained by planing the stone with a planer tool in which irregular nicks have been made in the cutting edge (Stone Catalog).

Obsidian. An ancient name for volcanic glass. Most obsidians are black, although red, green, and brown ones are known. They are often banded and normally have conchoidal fracture and a glassy luster. Most obsidians are rhyolitic in composition (AGI).

Onyx. 1. Translucent layers of calcite from cave deposits, often called Mexican Onyx or onyx marble.
2. A cryptocrystalline variety of quartz, made up of different colored layers, chiefly white, yellow, black (AGI).

Oolite. A spherical to ellipsoidal body, 0.25—2.00 mm. in diameter, which may or may not have a nucleus, and has concentric or radial structure or both. It is usually calcareous, but may be siliceous, hematitic, or of other composition (AGI).

Oolitic Limestone. A calcite-cemented calcareous stone formed of shell fragments, practically noncrystalline in character. It is found in massive deposits located almost entirely in Lawrence, Monroe and Owen Counties, Indiana and in Alabama, Kansas and Texas. This limestone is characteristically a freestone, without cleavage planes, possessing a remarkable uniformity of composition, texture and structure. It possesses a high internal elasticity, adapting itself without damage to extreme temperature changes (Stone Catalog).

Palletized. A system of stacking stone on wooden pallets. Stone which comes palletized is easily moved and transported by modern handling equipment. Palletized stone generally arrives at the job site in better condition than unpalletized material (Stone Catalog).

Pegmatite. Those igneous rocks of coarse grain found usually as dikes associated with a large mass of plutonic rock of finer grain size. The absolute grain size is of lesser consequence than the relative size. Unless specified otherwise, the name usually means granite pegmatites, although pegmatites having gross compositions similar to other rock types are known. Some pegmatites contain rare minerals rich in such elements as lithium, boron, fluorine, niobium, tantalum, uranium, and the rare earths (AGI).

pH. The negative logarithm of the hydrogen ion activity (less correctly concentration). For example, pH 7 indicates an H^+ activity of 10^{-7} mole/liter (AGI).

Phaneritic. A textural term applied to igneous rocks in which all the crystals of the essential minerals can be distinguished with the unaided eye. The adjective form phaneritic is currently used more frequently than the noun (AGI).

Phenocryst. A porphyritic crystal; one of the relatively large and ordinarily conspicuous crystals of the earliest generation in porphyritic igneous rocks. It has proved an extremely convenient term, although its etymology has been criticized (AGI).

Plucked Finish. Obtained by rough planing the surface of stone, breaking or plucking out small particles to give rough texture (Stone Catalog).

Plumose Markings. Feather-like markings on tension and extension joints. Also often found on split slate surfaces as networks resembling segments of spider webs, which start out from the chisel mark.

Pneumoconiosis. Dust diseases of the lungs. Includes silicosis.

Polished Finish. The finest and smoothest finish available in stone. Generally only possible on hard, dense material (Stone Catalog). A mirror-like glossy surface which brings out the full color and character of the marble. A polished finish is usually not recommended for marbles intended to be used on building exteriors (MIA).

Porphyritic. A textural term for those igneous rocks in which larger crystals (phenocrysts) are set in a finer groundmass which may be crystalline or glassy, or both (AGI).

Porphyry. A term first given to an altered variety of porphyrite on account of its purple color, and afterwards extended by common association to all rocks containing conspicuous phenocrysts in a fine-grained or aphanitic groundmass. The resulting texture is described as porphyritic. In its restricted usage, without qualification, the term porphyry usually implies a hypabyssal rock containing phenocrysts of alkali feldspar, though in the field it is generally allowed a wider scope, and commercially it is used for all porphyritic rocks (AGI).

Quarry. An open or surface working, usually for the extraction of building stone, as slate, limestone, etc. In its widest sense the term *mines* includes quarries, and has been sometimes so construed by the courts; but when the distinction is drawn, *mine* denotes underground workings and *quarry* denotes superficial workings (AGI).

Quarry Face. The freshly split face of ashlar, squared off for the joints only, as it comes from the quarry, and used especially for massive work (AGI).

Quartzite. A quartz rock derived from sandstone, composed dominantly of quartz, and characterized by such thorough induration, either through cementation with silica or through recrystallization, that it is essentially homogenous and breaks with vitreous surfaces that transect original grains and matrix or interstitial material with approximately equal ease. Such a stone possesses a very low degree of porosity and the broken surfaces are relatively smooth and vitreous as compared with the relatively high porosity and the dull, rough surfaces of sandstone (ASTM).

Redox Potential. 1. Same as oxidation-reduction potential. 2. The oxidation-reduction potential of an environment; in other words, the voltage obtainable between an inert electrode placed in the environment and a normal hydrogen electrode, regardless of the particular substances present in the environment (AGI).

Ribbon Slate. Parallel bands of variable thickness in slate which cut the slaty cleavage at an angle. Bands or ribbons are former bedding planes and may be lines of weakness; discoloring is also possible along the ribbons.

Rift. A planar property whereby granitic rocks split relatively easily in a direction other than the sheeting (AGI).

The most pronounced direction of splitting or cleavage of a stone. Rift and grain may be obscure, as in some granites, but are important in both quarrying and processing stone (Stone Catalog).

Rip-Rap. Irregularly shaped stones used for facing bridge abutments and fills. Stones thrown together without order to form a foundation or sustaining wall (Stone Catalog).

Rock Burst. A sudden and often violent failure of masses of rock in quarries, tunnels and mines (AGI).

Rock Face. This is similar to split face, except that the face of the stone is pitched to a given line and plane, producing a bold appearance rather than the comparatively straight face obtained in split face (Stone Catalog).

Roundness. The ratio of the average radius of curvature of the several corners or edges of a solid to the radius of curvature of the maximum inscribed sphere. Not to be confused with sphericity (AGI).

Rubble. A product term applied to dimension stone used for building purposes chiefly walls and foundations, and consisting of irregularly shaped pieces, partly trimmed or squared, generally with one split or finished face, and selected and specified within a size range (Stone Catalog).

Salt. Halite, common salt. Sodium chloride (NaCl). In chemistry, any class of compounds formed when the acid hydrogen of an acid is partly or wholly replaced by a metal or a metallike radical; as ferrous sulfate ($FeSO_4$) is an iron salt of sulfuric acid (AGI).

Sand. Detrital material of size range $1/16$—2 mm. diameter. Very coarse, 1—2 mm.; coarse, $1/2$—1 mm.; medium, $1/4$—$1/2$ mm.; fine, $1/8$—$1/4$ mm.; very fine, $1/16$ to $1/8$ mm. (AGI).

Sand-Sawn Finish. The surface left as the stone comes from the gang saw. Moderately smooth, granular surface varying with the texture and grade of stone (Stone Catalog).

Sandstone. A cemented or otherwise compacted detrital sediment composed predominantly of quartz grains, the size grades of the latter being those of sand. Mineralogical varieties such as feldspathic and glauconitic sandstones are recognized, and also argillaceous, siliceous, calcareous, ferruginous and other varieties according to the nature of the binding or cementing material (AGI).

A sedimentary rock consisting usually of quartz cemented with silica, iron oxide or calcium carbonate. Sandstone is durable, has a high crushing and tensile strength, and a wide range of colors and textures (Stone Catalog).

Sawed Face. A finish obtained from the process used in producing building stone. Varies in texture from smooth to rough and coincident with the type of materials used in sawing — characterized as diamond sawn; sand sawn; shot sawn (Stone Catalog).

Scale. Thin lamina or paper-like sheets of rock, often loose, and interrupting an otherwise smooth surface on stone (Stone Catalog).

Schist. A foliated metamorphic rock (recrystallized) characterized by thin folia that are composed predominantly of minerals of thin platy or prismatic habit and whose long dimensions are oriented in approximately parallel positions along the planes of foliation. Because of this foliated structure schists split readily along these planes and so possess a pronounced rock cleavage. The more common schists are composed of the micas and mica-like minerals (such as chlorite) and generally contain subordinate quartz and/or feldspar of comparatively fine-grained texture; all gradations exist between schist and gneiss (coarsely foliated feldspathic rocks) (Stone Catalog).

Scoria. Irregular masses of lava resembling clinker or slag; may be cellular (vesicular) dark colored and heavy (Stone Catalog).

Serpentine. A hydrous magnesium-silicate material of metamorphic origin, generally of very dark green color with markings of white, light green or black. One of the hardest varieties of natural building stone (Stone Catalog).

Serpentine Marble. A green marble characterized by a prominent amount of the mineral serpentine (MIA).

Shear. A type of stress; a body is in shear when it is subjected to a pair of equal forces which are opposite in direction and which act along parallel planes (Stone Catalog).

Sheeting. In a restricted sense, the gently dipping joints that are essentially parallel to the ground surface; they are more closely spaced near the surface and become progressively further apart with depth. Especially well developed in granitic rocks (AGI).

Shot Sawn Finish. A rough gang-saw finish produced by sawing with chilled-steel shot (Stone Catalog).

Silicate. A salt or ester of any of the silicic acids, real or hypothetical; a compound whose crystal lattice contains SiO_4 tetrahedra, either isolated or joined through one or more of the oxygen atoms to form groups, chains, sheets or three-dimensional structures (AGI).

Siliceous (also, Silicious). Of or pertaining to silica; containing silica, or partaking of its nature. Containing abundant quartz (AGI).

Silicosis. Disease of the human lungs. A condition of massive fibrosis of the lungs marked by shortness of breath and caused by prolonged inhalation of silica dusts (WEBSTER). See Pneumoconiosis.

Slate. A microgranular metamorphic rock derived from argillaceous sediments and characterized by excellent parallel cleavage entirely independent of original bedding, by which cleavage the rock may be split easily into relatively thin slabs (ASTM).

Soapstone. A massive variety of talc with a soapy or greasy feel, used for hearths, washtubs, table tops, carved ornaments, chemical laboratories, etc., known for its stain-proof qualities (Stone Catalog).

Soft Rock. Can be removed by air-operated hammers, but cannot be handled economically by pick (AGI).

Solubility. The maximum concentration of solute, at a given temperature and pressure, which can be obtained by stirring the solute in the solvent; also the concentration of the solute remaining when the temperature and pressure are changed to the given ones from values where the concentration of solute is higher. In other words, solubility is the equilibrium concentration of solute when undissolved solute is in contact with the solution. The most common units of solubility are grams of solute per 100 grams of solvent and moles per liter of solution (AGI).

Solubility Product. The equilibrium constant for the process of solution of a substance (generally in water). A high value indicates a more soluble material (AGI).

Solution. 1. The change of matter from the solid or gaseous state into the liquid state by its combination with a liquid; when unaccompanied by chemical change, it is called physical solution; otherwise, chemical solution.

2. The result of such change; a liquid combination of a liquid and a nonliquid substance (AGI).

Sorting. 1. In a genetic sense the term may be applied to the dynamic process by which material having some particular characteristic, such as similar size, shape, specific gravity, or hydraulic value, is selected from a larger heterogeneous mass.

2. In a discriptive sense, the term may be used to indicate the degree of similarity, in respect to some particular characteristic, of the component parts in a mass of material.

3. A measure of the spread of a distribution on either side of an average (AGI).

Spall. A stone fragment that has split or broken off (Stone Catalog).

Sphericity. The degree in which the shape of a fragment approaches the form of a phere. Compare angularity, roundness (AGI).

Split Face (Sawed Bed). Usually split face is sawed on the beds and is split either by hand or with machine so that the surface face of the stone exhibits the natural quarry texture (Stone Catalog).

Stone. 1. Concrete earthy or mineral matter. A small piece of rock. Rock or rock-like material for building. Large natural masses of stone are generally called rocks; small or quarried masses are called stone; and the finer kinds, gravel or sand (AGI).

2. A precious stone; a gemstone (AGI).

3. Sometimes synonymous with rock, but more properly applied to individual blocks, masses, or fragments taken from their original formation or considered for commercial use (Stone Catalog).

Strain. Deformation resulting from applied force; within elastic limits strain is proportional to stress (AGI).

Stratification. A structure produced by deposition of sediments in beds or in layers (strata), lamina, lenses, wedges, and other essentially tabulars units (AGI).

Stress. 1. Force that results in strain. 2. Resistance of a body to compressional, tensional or torsional force (AGI).

Stress Relief. Stresses dormant in deeply buried rock masses start to show phenomena of stress relief after erosion has removed overburden, such stress relief results in exfoliation, sheeting, rock bursts.

Strip Rubble. Generally speaking, strip rubble comes from a ledge quarry. The beds of the stone, while uniformly straight, are of the natural cleft as the stone is removed from the ledge, and then split by machine to approximately four inches width (Stone Catalog).

Stripping. To remove from a quarry, or other open working, the overlying earth and disintegrated or barren surface rock (AGI).

Strips. Long pieces of stone — usually low height ashlar courses where length to height ratio is maximum for the material used (Stone Catalog).

Structure. In petrology, one of the larger features of a rock mass, such as bedding, flow banding, jointing, cleavage, and brecciation; also the sum total of such features; contrasted with texture (AGI).

Stylolite. A term applied to parts of certain limestones and also to some other rock, caused by pressure-solution; the "columns", darker zig-zag lines than the surrounding rock, run usually roughly parallel to the bedding planes, but may also cut across bedding planes at angles to 90°. Concentrations of clay and/or organic substance give the stylolites (also called crowfeet) a distinct color and pattern (modified from AGI).

Syenite. A plutonic igneous rock consisting principally of alkali feldspar usually with one or more mafic minerals such as hornblende or biotite. The feldspar may be orthoclase, microcline or perthite. A small amount of plagioclase may be present. Also of quartz if less than 5%. Quartzfree granite. Name from Syene (Aswan), where it was later renamed "Aswan Red Granite" (modified from AGI).

Tectonic. Of, pertaining to, or designating the rock structure and external forms resulting from the deformation of the earth's crust (AGI).

Tectonic Breccia. An aggregation of angular coarse rocks formed as the result of tectonic movement. Included in this category are fault breccias, especially those associated with great overthrust sheets, and fold breccias (AGI).

Tectonic Conglomerate. A coarse clastic rock produced by deformation of brittle, closely jointed rocks. Rotation of the joint block and granulation and crushing sometimes produce a rock that closely simulates a normal conglomerate (AGI).

Terrazzo. A type of concrete in which chips or pieces of stone, usually marble, are mixed with cement and are ground to a flat surface, exposing the chips which take a high polish (Stone Catalog).

Texture. Geometrical aspects of the component particles of a rock, including size, shape, and arrangement (AGI).

Travertine. Calcium carbonate, of light color and usually concretionary and compact, deposited from solution in ground and surface waters. Extremely porous or cellular varieties are known as calcareous tufa, calcareous sinter, or spring deposit. Compact, banded varieties, capable of taking a polish are called onyx-marble. Travertine forms the stalactites and stalagmites of limestone caves, and the filling of some hot-spring conduits (AGI).

Tufa. Should not be confused with "tuff". See Travertine.

Tuff. A rock formed of compacted volcanic fragments, generally smaller than 4 mm. in diameter (AGI).

Veneer Stone. Any stone used as a decorative facing material which is not meant to be load bearing (Stone Catalog).

Verde Antique. A dark-green rock composed essentially of serpentine (hydrous magnesium silicate). Usually criss-crossed with white veinlets of magnesium and calcium carbonates. Used as an ornamental stone. In commerce often classed as a marble (AGI).

Volcanic Block. A subangular, angular, round, or irregularly shaped mass of lava, varying in size up to several feet or yards in diameter (AGI).

Volcanic Breccia. 1. A more or less indurated pyroclastic rock consisting chiefly of accessory and accidental angular ejecta 32 mm. or more in diameter lying in a fine tuff matrix. If the matrix is abundant, the term tuff breccia seems appropriate.

2. Rock composed mainly of angular volcanic fragments of either pyroclastic or detrital origin coarser than 2 mm. in a matrix of any composition or texture or with no matrix.

3. Rock composed of angular nonvolcanic fragments enclosed in a volcanic matrix (AGI).

Vug. 1. A cavity, often with a mineral lining of different composition from that of the surrounding rock.

2. A cavity in the rock, usually lined with a crystalline incrustation (AGI).

Weathering. The group of processes, such as the chemical action of air and rain water and of plants and bacteria and the mechanical action of changes of temperature, whereby rocks on exposure to the weather change in character, decay, and finally crumble into soil (AGI).

Wedging. Splitting of stone by driving wedges into planes of weakness (Stone Catalog).

Wire Saw. A method of cutting stone by passing a twisted, multi-strand wire over the stone, and immersing the wire in a slurry of abrasive material (Stone Catalog).

Xenolith. A term applied to allothigenous rock fragments that are foreign to the body of igneous rock in which they occur. An inclusion (AGI).

Bibliography of Sources for Glossary

Glossary of Geology and Related Sciences. — American Geological Institute, 2nd ed., 1960. Abbrev.: (AGI).

Marble Engineering Handbook, Appendix E, Definitions. — Marble Institute of America, Inc. Washington, D.C., AIA File No. 8 B-1, 1962. Abbrev.: (MIA).

Standard Definitions of Terms Relating to Natural Building Stone. — Am. Soc. Testing Materials, ASTM Designation: C 119—50, 1958. Abbrev.: (ASTM).

Stone Catalog, 1968/69. Building Stone Institute, Glossary of Terms. — New York. Abbrev.: (Stone Catalog).

Index of Names

Subject Index

Bold Face Indicates Major Entry

Printer: R. Spies & Co., A-1050 Wien